CHARLESTON BLACKSMITH

CHARLESTON BLACKSMITH
The Work of Philip Simmons

John Michael Vlach

The University of Georgia Press

Athens

COPYRIGHT © 1981 BY
THE UNIVERSITY OF GEORGIA PRESS
Athens, Georgia 30602

Set in Century Schoolbook and Korinna types
Printed in the United States of America
Design by Gary Gore

Library of Congress Cataloging in Publication Data

Vlach, John Michael, 1948–
  Charleston blacksmith.

  Includes bibliographic references.
    1. Simmons, Philip. 2. Blacksmiths—South Carolina—
Charleston—Biography. 3. Charleston, S.C.—Industries. 4.
Charleston, S.C.—Biography. I. Title.

TT220.S55V52   739'.4724 [B]   80–16632
ISBN 0–8203–0533–2
ISBN 0–8203–0586–3 pbk.

FOR DANIEL J. CROWLEY

CONTENTS

ILLUSTRATIONS

PREFACE

WHEN I first saw him, he was bent over a pile of odds and ends searching for a useful piece of iron. Nervous about this initial meeting, I cleared my throat and said, "You must be Philip Simmons." Straightening up, he looked quizzically at me, and then, smiling widely, offered his hand in greeting.

That was in 1972, when I was a graduate student at Indiana University. Since then I have made many visits to Philip's shop for varying lengths of time. Sometimes my purposes were tied to academic ventures; other times we simply chatted and exchanged gripes and hopes. I am sure I amused him. Here I was, a young white stranger who came hundreds of miles across the country to ask him about the old ways of black people in Charleston. Curious about the life and work he took for granted, I must have seemed odd to him, to say the least. Nevertheless, it made sense to him that I asked my questions, because he had knowledge to give and he wanted to share it. Philip liked the old ways of ironwork, which he had helped preserve, and because I, a folklorist, had an avid interest in those ancient customs and practices, he came to like me too. Caring about the same things, we became friends.

This book is partial repayment of a debt of friendship; I owe Philip Simmons for his patience, understanding, and openness. Simmons is a teacher: he gave me lessons about blacksmithing and lessons about living. The elders of any age are wise, but, more important, they give the confidence to persevere by the example of their existence.

The account that follows has been pieced together from several sources. I have made extensive use of my own field notes, a set of random jottings and observations made from 1972 to 1978. In 1976 and again in 1978 I interviewed Philip on tape. Excerpts from these fifteen or so hours of recorded conversation constitute the major source of the narrative presented in this book. I also interviewed Lonnie Simmons, one of Philip's apprentice mates and the son of Philip's teacher; James Kidd, another Charleston blacksmith, now retired; and Willie Williams, a current apprentice. Other tape-recorded material was collected for the African Diaspora section of the Smithsonian Institution's Folklife Program in 1976 and 1977 by William Wiggins, James Early, and Jason Dotson. I have dovetailed their interviews with Philip and two of his younger partners, Silas Sessions and Ronnie Pringle, into the longer narrative that I collected. Other sources of information included a videotape of Philip by a South Carolina educational television program for children called "Studio See," the clipping file of the *Charleston News and Courier*, and a few brief magazine articles.

None of the previous discussions of Philip's career has ever given a complete account of his life from beginning to end. Indeed, in our conversations Philip's story emerged erratically. He might, for example, recall his first job as a blacksmith, then remember the work of a former colleague, and end with a reminiscence about baseball games or fishing. His ability to recall the events of his long life was also uneven. The distant past was crystallized into distinct stories; events which occurred between 1945 and 1970 were murky in his mind. Consequently, some sequences are richly detailed and others are vague. I have welded all of the data I could gather into a chronological chain. Where breaks occurred I used the historical literature on Charleston and the surrounding area, in addition to studies of blacksmithing and ornamental wrought iron, to forge a reasonable link. Because our interviews strayed widely and randomly over many topics, we sometimes covered the same ground more than once. In such intances the second or third interview might produce an interesting detail that could be added to the basic story. In editing Philip's words, I have united some of these diffuse bits of talk, which sometimes occurred years apart, as if they were all part of one continuous conversation. While this approach is ethnographically inaccurate, it does contribute to the formulation of a well-rounded and better-informed account. Moreover, because all that Philip has told me is part of a single life history, I do not think my editorial manipulation too disruptive of the facts.

All of the passages that are set in quotation marks have been transcribed accurately from tapes, field notes, newspaper articles, and other sources. I have occasionally had to worry about punctuation or to introduce bracketed phrases in order to make a verbatim text readable and comprehensible. This is always a problem with oral testimony, where much information is conveyed by visage, gesture, intonation, and physical context. But beyond these modifications, I have not tampered with the words. Some may wonder about the inconsistency of a text where, for example, the word *nothing* is alternately spelled with and without the final *g*. This was done because of Philip's use of alternate pronunciations. The reader will encounter many other colloquialisms. In my fidelity to Philip's words as spoken I have also retained unconventional grammatical formations. I hope that no reader will sense any intellectual snobbery or exploitation on my part. These linguistic constructions are accepted as legitimate and valid in black English, and I thought that any attempts to "correct" Philip's ethnic speech pattern would be pompous and presumptuous.

This book is a professional biography; it deals mainly with a craftsman's work. Consequently the central focus is blacksmithing. Other dimensions of Philip's life are included but are considered here as minor themes. To establish a cultural context for his career, the first chapter gives the details of his childhood in the Sea Islands, where he acquired, among other things, his work habits and the desire always to remain busy. The second chapter moves more directly into his apprenticeship as an ironworker and is followed by a discussion of his maturation in the craft. This third section describes some of the work Philip has done and contains a catalog of the major decorative works upon which his fame is based. The fourth chapter suspends chronology in order to review in detail Philip's attitudes about his decorative work; it contains a discussion of his ideas about art and aesthetics. In chapter five an attempt is made to round out the reader's view of the man. I review here some of his various roles in his community: family patriarch, church leader, entrepreneur, and baseball player. Having considered his wider importance to the community, I then return to the blacksmithing profession's influence on the community. This influence is realized chiefly through the training of apprentices, passing the tradition on to the next generation. The experience of Philip's most recent trainee, Willie Williams, allows us some understanding of both the ability of the teacher and the enthusiasm of the student. In the concluding chapter, Philip gives his sense of the future of his craft in Charleston.

I hope that with this combination of chronological reconstruction and thematic inquiry I have rendered a comprehensive statement of the life and work of a blacksmith. I have stressed the human component of the craft, leaving aside many of the technological complexities. Instead of explaining the processes of metalworking, I have presented the story of one craftsman's struggle to keep the past alive in the twentieth century. While only a specialist can really understand technical descriptions, almost any reader will appreciate this saga of survival.

To allow the interested reader to consider in more depth the issues raised by this biography, I have provided a bibliographical discussion which cites other works of interest. Also noted are the historical sources that I have used to hold separate elements of this book together. In the third chapter I could not avoid employing a number of specialized terms used in the description of decorative wrought iron. A glossary of these and other technical terms has been provided. Since this book is the record of a person's life, its focus is historical. Many details about blacksmithing

enter this account, but technical information has to a great extent been avoided. For those who want to understand how wrought-iron objects are made I have included an appendix which illustrates the making of a tool and a fancy piece of ornamental wrought iron.

Two current books have influenced this work: *The Handmade Object and Its Maker* by Michael Owen Jones, and *All Silver and No Brass* by Henry Glassie. While Jones writes about a Kentucky chairmaker and Glassie describes Christmas mumming plays, both present in detail the individual bearers of tradition. It was this rich and extensive treatment of the individual that I liked. In that same vein, I have tried to let Philip's character emerge here. I hope that my use of transcribed material will allow the reader to feel as if he himself has engaged Philip in a long conversation. The illustrations obviously intertwine with the verbal account, but they also run parallel to it. They can be taken by themselves as a record of artistic achievement. Each piece of work can tell its own story to some degree if one is sensitive to form, shape, line, and texture. Thus the book combines the approach of the oral historian with that of the art historian. This synthesis of academic methods is typical of current folklife scholarship. We find that we must breach disciplinary walls if we are to tell important tales that have been overlooked before.

This book has been a long time in the making. I was first asked to report on Philip's work by Mary Twining in 1973, for the benefit of the American Folklore Society. I made a brief presentation, and subsequently William Ferris prodded me to prepare an article on my research; that article was published in 1976.

Philip Simmons liked the piece (particularly the small sketches of the stages of leaf making) and agreed that a book might be interesting to see. With Philip's cooperation, then, the article grew into this book.

While gathering material I visited Charleston and Philip often. Those visits were profitable in large measure because of the generosity of my aunt and uncle, Rosemary and John Bowman, who also live in Charleston and who provided me with a home away from home. Finally, in the summer of 1978, when I began in earnest to bring all the loose ends together for this book, I discovered that I needed to take many more photographs and get many questions answered. The University Research Institute of the University of Texas awarded me a generous grant which allowed me to do two months of uninterrupted fieldwork. My friends and academic colleagues Archie Green and Howard Wight Marshall both read the manuscript while still in the rough stages and provided many helpful suggestions that improved the final draft. Beverly Benadom typed the manuscript, cheerfully and efficiently showing by her enthusiasm that other readers might also enjoy this book. Finally, Ellen Harris gave the book a good once-over while editing the final copy. I would like to take this opportunity to thank the people and institutions who have given me both moral and financial support in this project.

One day in June 1978, Philip introduced me to a customer who had come into his shop. He explained that I had come from Texas, and that I was writing a book about him. The customer was startled to think that a local ordinary citizen would attract visitors from such a distance, and that this familiar person would be the subject of a book. Philip chuckled a bit and responded, "Well, I guess there ain't too many people got books about them. Just George Washington, Abraham Lincoln, and me." In our society we usually enshrine great people in books. This book is but one brick in a shrine; the rest of the monument Philip has already made for himself with a life of ironwork.

CHARLESTON BLACKSMITH

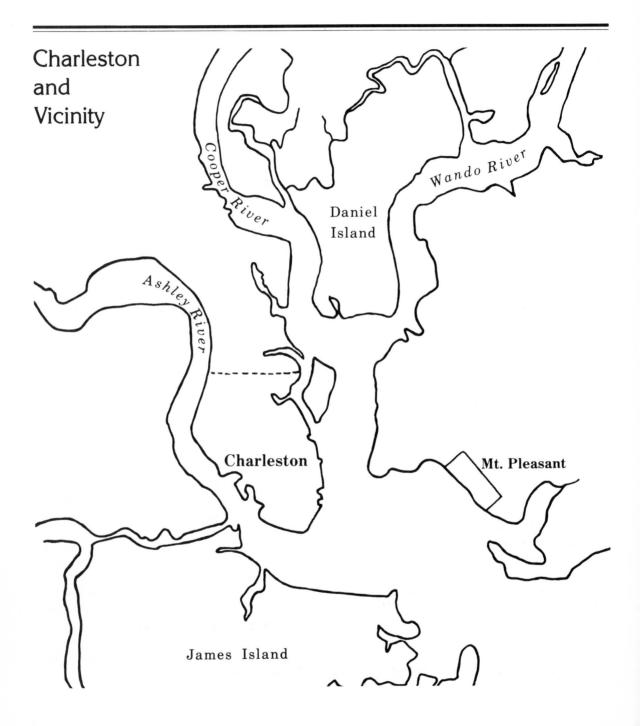

Charleston
and
Vicinity

Cooper River

Wando River

Daniel
Island

Ashley River

Charleston

Mt. Pleasant

James Island

1

"Poor People Done It Like That"

ACROSS the Cooper River from Charleston there stands on the shore a dense thicket of trees. This forest is interrupted by stretches of lush green marsh grass. The vast foliage of leaf and blade gives the impression that a broad expanse of land lies beyond. But it is just an impression, for what you see is only Daniel Island, a small dot of soil in the South Carolina low country. Surrounded by the Cooper and Wando rivers and a series of interlinked swamps and creeks, it is technically an island. At the threshold of one of the oldest cities in the United States, the island's watery boundaries have kept it astonishingly isolated. Even now only one very roundabout road provides access. Daniel Island was once plantation land owned by Robert Daniel of Barbados, a colonial governor of South Carolina, and it has remained farm land ever since the end of the seventeenth century. It was here on a small holding, in 1912, that Philip Simmons was born. Destined to become a blacksmith, he was first an islander who learned the ways of the farm and the river. For more than half a century he has lived in Charleston and could rightly claim the cosmopolitan character of that city. Yet when asked about his origins, he is quick to say:

I came from Wando, Daniel's Island. That's is in Wando and Wando is in Berkeley County. I born in Wando, up there in Wando. That's just near Cainhoy.

Philip led the life of a Sea Island farmer and fisherman for his first thirteen years. He stayed on the island constantly until he was eight years old. It was then that he first saw Charleston.

At that particular time children, student in school, then they [didn't] start school until they were seven years old. Now it is six but then it was seven. My seventh year—birthday—I went to school, start to school. I went to school one year in Berkeley County and the next year the teacher didn't come. The money—the budget—they didn't get up the budget for 'em. At that time it was a little slow gettin' that money. Money was scarce so the teacher didn't come early that year. They came about two months later. So my granddaddy send me here to Charleston.

But when summer came he returned to the island.

Used to work with my granddaddy in the country, fish and farming. I made a living off that. I helped buy all my clothes to come to school to Charleston. When I come to Charleston to go to school in September I got all my clothes and everything I need. During the summer months I work and I made enough money. . . . Let me say every year until I was thirteen, I went back in the country and farm with my grandparents and fish.

So every day for his first eight years, and during summer for the next five, Philip

1

acquired the rural values of a Sea Islander. His great-grandparents had once been slaves, and they told him about those days in bondage. He learned that the work was hard then and learned from his grandparents that conditions had remained hard. He came to know and prefer a busy life full of chores and tasks, a routine that he has never abandoned. It was his Sea Island upbringing that helped establish his current motto: "I like action."

The house in which Philip was born was fairly typical for the black residents of the region. It was a small frame dwelling with a lean-to shed across the rear; the whole building contained four rooms. There was an attic above that could have been used for a sleeping room, but the space was only used for storage. The house was sided with handriven clapboards. This dwelling, while quite ordinary, was particularly important to Philip because his grandfather had built it.

Our house was made out hand clapboard you get out of the woods. No mill work at all attached to it. My grandfather built it. He was everything. He was a poor man. He wasn't a professional. He built his own house, wasn't a professional house builder. He wasn't a professional fisherman. He wasn't a professional farmer. But that's the thing that he done.

While Philip admired the effort and resourcefulness of the builder, we might further marvel at the continuity of architectural tradition. Those tiny frame houses were more than an element of the Anglo-American plantation era. They were survivals of an Afro-Caribbean plan first used in Barbados, the immediate homeland of the first black settlers in the Sea Islands. Houses of this type were built in the general area until very recently. We may only conjecture at this point on the importance of Sea Island architecture on Philip's upbringing, but consider the scenes of family life in close quarters. One is never far away from anyone; one hears most of what goes on; one's sense of self is secondary to one's sense of family or group. We might conclude that an individual's character will be profoundly shaped by such circumstances, and we will eventually see that some aspects of Philip's personality, particularly his devotion to his family and the members of his church, can perhaps be traced back to the small clapboard house on Daniel Island.

The larger room in the house was sparsely furnished with a table and a few chairs that Philip's grandfather made. It was a lean existence, a life of poverty but not debilitating poverty. His folks didn't have much money, but they were skillful and made do well enough.

Back then they didn't have a way to cook . . . especially the poor ones. They cook on fire. The rich people, they use these different [fireplace] cranes and stuff. That was modern for them at that time. But we had to go 'bout it the old way. Right down there, put the potatoes in the ashes. And you could bake potato, man! Take it off and brush the ashes and once you peel it you think it was cook in the oven. Rich people had those cranes. Poor people couldn't afford it. Had to put bricks on the two sides. Couldn't afford dog iron. Bricks on two sides. Put your stuff on there. You burn the ashes, you burn the coals. You burn the wood while he green, and after it burn a certain time, he get red and it coal. Then you put your pot on there. Brother it cook right. Oh, it cook nice! Then you put some that coals on the side and poor people done it like that. That's the way my grandmother done it.

Philip cherishes the hours spent in his grandmother's kitchen. He says, "And you see I glad for that experience."

Philip's memories of his "Dan's Island" days are of course vague for the first six years: "You can't remember nothin' much before the age of six anyhow." But we can piece together a broad outline of daily life in the Sea Islands which would reasonably approximate his experience. His grandfather was a farmer and had to match his actions to the shifts in climate and season. There was spring planting, summer weeding, fall harvest, and winter "lay by." As a tiny lad, Philip followed his grandfather about and watched him perform his usual chores. Maybe Philip even got in the way. He remembers that his grandfather raised "all kinds of vegetables like corn, beans, and cotton—he planted a lot of cotton. Especially he'd plant stuff for the household. He'd plant enough for the market. That's the only way we had to live."

The farm was small, not more than thirty acres, as was the common Sea Island pattern. The farmstead was thus mainly an oversized kitchen garden intended to produce the staples for a year's self-sufficiency. Extra truck crops, if there were any, could be sold in the Charleston market. These were beans and cucumbers. Animals were kept too: usually chickens, a hog or two, and a solitary milk cow. During the first decade of the twentieth century, draft animals—oxen, mules, or horses—were used to pull plows, but only sporadically. Deep turning of the sandy soil was just then being introduced into the Sea Islands. This innovation in tillage did not completely displace the tradition of hoe cultivation, the method of their African forebears, which the islanders retained from plantation days. They used heavy hoes with broad blades hafted on thick saplings. Those hoes weren't at all delicate like our modern garden hoes. They were tools to give shape to the land, the prods that stirred the tired soil into producing one more year's sustinence. Heavy, rough, and crude, they were not for children to use. No doubt Philip watched his grandparents from the edge of the field as they raised their greens, potatoes, corn, beans, and rice.

Philip's grandmother was no stranger to field work. Like most Sea Island women, she was not housebound and consequently took a place in the garden patch along with her husband. With a bandana around her head and dressed in a loose-fitting blouse with the sleeves rolled up and a long skirt tied with a string at the hips, she toiled both in the fields and the kitchen. Thus it was that Philip had two models for action. When he was a child, everyone worked all the time and worked hard at the same jobs. For the rest of his life he would know and prefer no other way of living than working full time, all the time.

One of Philip's first memories of farming tasks involved the use of a mortar and pestle (plate 1). The mortar was a section of an oak trunk standing about three feet high and measuring fifteen inches across. Philip's grandfather had hollowed out a section of the top with his axe and by controlled burning. At age six Philip was too short to work at the

*1   Threshing rice with a
mortar and pestle.*

mortar. He had to stand on top of a box to look inside, and he wasn't strong enough to lift the regular pestle. Nevertheless, he was allowed to play at "mashin' the corn." The pestle, a wooden pole with two blunt ends, was too big for Philip's little hands, so his grandfather made a small one for him to use. When his grandmother prepared corn or rice, Philip could then take part in the work. He made his contribution, even though it certainly couldn't be regarded as a full share. The cooperative use of two pestles was not uncommon in the Sea Islands; the husking of rice was often done by two people working in unison.

As he became older, he abandoned his small pestle. Philip regarded it as a toy, a plaything. He threw it away without regret when he found that he could master the implement his grandparents regularly used. Not only was it

bigger, it was a more specialized tool. One end was broad and blunted into a dull roundness through its many collisions with the mortar. The other end was kept wedge-shaped like a broad chisel. This pestle was a proper rice-husking tool. While the Simmonses' farm had little acreage given to rice, rice was a major staple in the diet. It had been an element of the regional cuisine for the low country since 1690. The blunt end of the pestle was used for cracked rice, since it was the smashing end; the pointed end produced whole grain rice. In either case, the husk of the rice was separated from its kernel. Once the husking was completed, the two parts of the rice grain lay jumbled in the mortar and further cleaning was required. This phase was done with a round, traylike, coiled-grass basket called a fanner. The contents of the mortar were placed in the fanner and tossed gently into the air, allowing the lighter husks to blow away. Philip claims never to have "flapped the rice," but he saw it done many times.

Reaching the age of seven, Philip started to school. It was a one-room school for the black children on Daniel Island and, as he remembers, it only lasted for three months—just enough time to learn the alphabet, some numbers, and how to write his name. This left most of the year free to do farm work. Many of his chores involved carrying and hauling. While he was never charged with milking the cow, he certainly had to carry its feed and at times take the milk pail to the house. His grandmother usually fed the chickens, but he would do that too. Occasionally it was his job to carry slops to the hogs, and it always was his job to gather up the firewood.

At the age of eight, Philip was given the task of carrying "dinner" to the working people in the fields. The annual field cycle first involved enriching the soil, usually following the local practice of spreading a natural compost of marsh mud mixed with manure and working it into the ground with hoes. Planting would be finished by April, and throughout the spring and summer weeding required constant vigilance. As the bean plants matured they were picked as needed and the small crop taken to the house for supper. The other vegetables—the peas, tomatoes, collards, and okra—were harvested in their turn. With the coming of fall, the potatoes were dug up and banked near the house to be used through the winter. There were certain slack periods in the growing

process when the field did not require immediate attention, but there was never a time when there was nothing to do. The household chores were constant and tended to fill the momentary lulls in Philip's routine.

Beyond the tasks of field and house, Philip also gave a hand to his grandfather's carpentry. William Simmons's skills as a joiner were apparently valued in the Daniel Island community. He built his own house; he may have built houses for others. If he could build a house, it is likely that he also built barns, sheds, smokehouses, and other structures. Such buildings were part of the architectural repertoire of Daniel Island, and William Simmons may have been one of the carpenters responsible for their construction. What Philip remembers best, however, is that his grandfather made caskets. These were fashioned with milled pine boards and formed into an elongated six-sided box. The angles were tricky to figure, and the joinery had to be precise. The importance of funerary rites in the Sea Island tradition required that the remains of the deceased be lavished with as much ceremony and finery as possible. To give anything less than one's best effort would be to invite the fury of the spirit of the deceased person. The wooden caskets were simple and unadorned, yet they were a fine display of the country carpenter's skill. Philip helped to steady the boards as they were sawn. He searched for the tool that his grandfather called for. He braced his weight against the sides as they were nailed or screwed together. In short, he lent a hand and learned some basic lessons about making objects, about attention to detail, about the responsibility of the craftsman to his client.

When there was no demand for carpentry and the fields could be left untended (particularly after the potatoes had been dug up), William Simmons became a fisherman. Indeed, throughout the entire Sea Island chain running from Georgetown, South Carolina, to northern Florida, fishing in the rivers and creeks has always served as a means of supplementing the diet. Myriad species of water creatures are caught, including bass, perch, trout, salmon, flounder, croaker, shrimp, clams, oysters, crabs, and turtles. On Daniel Island there were two main types of fishing, line fishing and shrimping. Speaking of his grandfather, Philip recounts this pattern:

He farmed part of the season, and the other part after harvesting, then he fished . . . he done shrimp. He catch shrimp for a living, fish for a living. Bring it to the market here [in Charleston]. You see where the main city market is over here; it's the same market we brought out produce like fish and shrimp. . . . We fish with line and net. Most times we'd fish for pleasure. We'd use the line. Commercial fishing for the market we'd use the net. Cast net we'd call it. Cast net and fish line. We cast net one tide, we'd sit down and we'd fish the other tide. Well, when it don't run in the net, we use the fishing line. We didn't use the fish pole and cane then, rod and reel. Weren't no rod and reels then.

It was only natural that a farmer from Daniel Island would bring shrimp to Charleston. The city is clearly visible from the mouth of the Wando River, which was the major fishing area. A good hour's row across the Cooper offered a reward of much-needed cash from hungry Charlestonians, who have long considered the rice and shrimp concoction they call "shrimps creole" an indispensable part of their cuisine.

The shrimping business is now highly industrialized, with trawlers grabbing virtually all of the coastal catch. Yet in the creeks and shallows amid the islands, shrimp can still be caught with the centuries-old cast net. This type of net is known both in Africa and the Caribbean, and Sea Islanders have fished with it since the eighteenth century. The net is conical in shape, with its outer edge or "foot" weighted with lead shot "bullets." Draw lines run from the foot to the center where they are spliced into a hand line. When the net is thrown, it unfurls to its full diameter over the water and then sinks, trapping the shrimp below it. There are several techniques for throwing the net. In one approach a portion of the net is held in the mouth to help spread the net quickly to its fullest dimensions as it is thrown. Precise hand-mouth coordination is required for a proper release. Philip comments:

I can throw it, yes I can throw it. . . . Some people put it in he mouth and some don't. Those expert don't put 'em in they mouth. They could throw it without puttin' it in their mouth. Hand line in one hand and net in the other. Throw it and spin it. Yeah, it's an art in doin' it. You throw it right, hold it right, throw it right, it'll open.

Philip learned to throw a cast net from his grandfather, who could also make them. While Philip never learned to "knit net," he did discover that there was a native taxonomy which varied with the size of the mesh. The larger openings identified a "rich man's net."

The rationale for this name was that the larger openings allowed some of the small shrimp to escape and "rich folk" didn't want to go through the trouble of cleaning the morsel-sized shrimp anyway. The poor man, on the other hand, with many mouths to feed, must capture as much as he can with every cast. Hence, his net has a fine mesh through which no shrimp can escape.

On a subsistence farm nearly everything produced is consumed. When the Sea Island farmer turned to the waterways, the same attitude prevailed. Industriousness in both lines of work could, however, provide a surplus, as it did for William Simmons. Extra vegetables and extra shrimp could be sold in the Charleston market. With the money gained he was able to buy a wood-burning stove, which transformed the food-preparation customs from those of the plantation slave to those of a freedman. This change occurred around 1919, half a century after slavery. At this time he also thought it wise that his grandson get a good education. It was clear to him that the brief and interrupted appearances made by the teacher at Daniel Island were not enough. Thus, in the fall of 1920, Philip was sent to Charleston to live with his parents and to go to school for eight months of the year—a proper education.

Philip was rowed over to Charleston and disembarked at the foot of Calhoun Street, a docking point for many Sea Island boats or *bateaux* as they were called. From there it was only a few blocks to the house on Vernon Street where Philip would live for the next three years. The move to the city was exciting for Philip. Imagine the vast difference between a city like Charleston, teeming with people, and the few clusters of tiny farm houses on Daniel Island. In Charleston there were paved roads, sidewalks, parks, multistory houses, office buildings made of stone, and ships at the wharves—and there were thousands and thousands of people. All of them were new and strange to Philip. On the island everyone had a clear identity; in Charleston the procession of strangers was endless. Adventure thus lay around every corner, particularly in the area surrounding Philip's house. Charleston was a bustling port city, and Philip was but two blocks from the waterfront where, in the 1920s, one could still see three-masted schooners and all manner of steamships. Wagons pulled by horses were still the order of the day, and drivers constantly urged their burdened teams back and forth between the docks and the warehouses. Many service trades were also located along the river, and Philip would spend hours with the cobbler, the pipefitter, the shipwright, the cooper, and other craftsmen.

Exciting though the city was, it remained only a temporary home. Philip was still a country boy and may have been prone to follow his grandmother's warning, "There time to see and not see, time to know and not know," which promoted a cautious, reticent view of life. He was happy when summer came and he could return to Daniel Island. Until about the age of eleven he was most comfortable with the regimen of the fields. The quest for shrimp and sea bass he saw as sport. Yet both endeavors were an economic necessity. The money Philip earned in the summer helped get him through the school year. During the rest of the year, in Charleston, he did odd jobs to supplement the money his mother made as a domestic helper. He shined shoes or sold newspapers, but these jobs were not important enough to keep him off the island.

Between the ages of eleven and thirteen Philip's role on the farm and in the boat increased tremendously. He was under less supervision and was expected to perform more responsibly, almost at the level of an adult. He worked the same hours as his grandfather, getting up at dawn and returning to the house at dusk. Philip's main memories are of the rush to harvest Irish potatoes before he returned to Charleston for school. Rows of plants had to be turned in order to unearth the knotty clumps of potatoes. Then Philip had to gather up the potatoes, shake the dirt off them, and carry them back to the house, where they would be stored. He took over other crops too: the corn, the sweet potatoes, and the beans. He learned more and more about plant cultivation and animal care, and if he had continued to return to Daniel Island beyond his thirteenth summer, he might have remained a farmer.

The impact of Philip's Sea Island childhood was particularly significant in terms of his sense of ethnic and regional identity. While he has lived and worked in the black community of Charleston most of his life, his first seven years were spent almost exclusively in the presence of blacks. White people rarely came to Daniel Island and then were seen only at the store in Cainhoy. Other than in trips to the Charleston market, Philip lived among immediate family, lesser kin and familiar neighbors who were all farmers as well as members of the same race and social class. They shared the same island setting; all knew nature intimately and had the same concerns produced by the fickleness of weather or shifts in the season. Foodways were so uniform from one household to another that all ate the same basic diet, which, by Philip's account, tended to concentrate on grits and rice. Moreover, on the basis of food-preservation customs, he can separate the residents of Yonges, Wadmalaw, and Daniel Islands; on other islands, "they canned," while on his island "they dried" their foodstuffs. His favorite dish remains one that his grandmother made, okra soup with dried shrimp in it. This should then be poured on top of a plateful of rice cooked "Geechee style, so dry that each grain fall apart."

It is the widely shared experience of the islanders and specific eating habits that cause Philip to identify himself first as a Geechee, a local term for a Sea Islander:

Well, Geechee [people] mostly like rice and most of the people come in see you eating a lot of rice, they call you Geechee. So I'm a colored, black Geechee.

In the social context of Charleston, which has long maintained a distinct black subculture of its own tied to a history of urban servitude, manumission, and the independence of free black tradesmen, to identify oneself as Geechee is to confirm one's status as an outsider. But Philip feels no shame or anxiety about this circumstance. As we have noted before, he is "glad for that experience." He considers rural life a good training with many hardships that have made him appreciate more fully the successes he has earned in Charleston.

But even more than identity, his island days provided links to some of the oldest elements in the black culture of the Carolina low country. One very important aspect of Sea Island tradition is found in his speech pattern.

Philip's speech is profoundly influenced by the rhythms and sounds of Gullah, the African-based creole spoken in the islands. Over the years since he left Daniel Island the Charleston version of black English and his deliberate attempts to learn "proper English" in school have modified his usage pattern, but the inflectional aspects of Gullah remain. Philip acknowledges that "that ol' Geechee" comes out now and again when his sentences tend to rise tonally for emphasis, what he calls "gainin' up." Creolists would no doubt perceive certain distinct Gullah grammatical forms such as unconventional pronoun usage or sentences with serial verbs, but it is enough for the present to recognize that after an absence of fifty years from Daniel Island Philip still speaks a form of Gullah.

The things he confronted daily on his grandparents' farm—the mortar and pestle, the coiled-grass baskets, the cast nets, the house itself—constitute a physical statement of tradition as old as Sea Island speech. Each was used in the plantation era and can be traced historically to an African genesis. Philip was, however, not conscious of this heritage; for him the immediate history of his grandparents and great-grandparents was more significant. He, in fact, finds it exotically interesting that elements of his life can be traced back to Africa. His usual comment on such matters is, "Well, poor people had to do it that way." It is nonetheless important that there were Africanisms in his early experiences, for they can be interpreted as a prime statement of the cultural conservatism of his background. The artifacts which surrounded him were old in form and concept. Philip today places high value on items made "the old-fashioned way." There can be no

denial that from his first days old ways were considered the best ways.

Certain patterns of behavior with which he was in daily contact are also traceable to Africa. In some cases the yard around the house was treated as an extension of the building. The dirt was swept regularly as if it were floor space. The soil was packed hard and no grass was allowed to grow, in contrast to the usual European pattern. Work was often done communally; tasks were shared when possible. Whether weeding the cotton, digging the potatoes, or smashing the corn in the mortar, the islanders preferred not to work alone. Cultivation was generally done with a hoe, not just because folks were too poor to afford traction animals: if hoes were used more people could join in (plate 2). Chanting as they worked, a monotonous job became more enjoyable.

If patterns in residence and labor are not enough to signal a sense of alternative cultural order, consider the habit of carrying loads on one's head. This practice on the part of both men and women is seen occasionally today in the low country. In the 1920s and 1930s it was so common as to be unremarkable. Since no one has reported even a single white farm wife carrying her supper greens back to the house on her head, we must consider the practice another source of black Sea Island identity. Having lived with all of these behavioral traditions, Philip would forever have a clear sense of his roots. The way of his grandparents was definitely the Sea Island way, the black way, the old-fashioned way. The links even to an African past are there too, even if not consciously acknowledged.

In the 1920s and 1930s many black people

*2   Communal work at
Hampton Plantation.*

left their Sea Island homes to go to cities all along the Atlantic coast. Many went to New York. Relatively few came to Charleston, mainly because there were few well-paying jobs. Philip had been migrating back and forth across the Cooper River for six years when he became a permanent Charleston resident. Intrigued by ironwork, he decided to become a blacksmith in the city, but he would never shake his country origins. The influences of the islands even today shape his point of view. With that in mind we can well imagine how much of an islander he was when he first stood at the door of Peter Simmons's blacksmith shop.

2

"I Have to Catch It As It Go"

IN SOME WAYS it was inevitable that Philip would become a blacksmith. As a boy he was fascinated by wrought-iron design, and living in Charleston, which is filled with examples, only fueled that interest. At the age of eleven he moved further up the peninsula to a frame "single house" on Washington Street. (See map, p. 106.) It was a short walk to Buist School, just two blocks down Washington and three over on Calhoun, but Philip followed a meandering path that took him over to East Bay Street past some elaborate "planters' houses." He turned the corner at Charlotte Street to gaze at the iron fence and large gate at number 16. From there he could either continue on along Charlotte or cut down Alexander or Elizabeth streets to Calhoun and scamper into the school yard just ahead of the bell. His own neighborhood was spare and plain without any decorative ironwork, but on his daily trek to school he peered into the world of the rich and near rich, which was profusely decorated. When Philip's teachers allowed time for the art lesson, he would sketch the ironwork that he had passed earlier in the morning:

Tell you what get me eager to draw, because I seen so much ironwork around Charleston—the early craftsmen and I'd like it.

That's what get me to start drawing and I do my drawing.

His attention already fastened on the work of the blacksmith, Philip eventually found his way to the smithy itself. It was, however, to take several years before he would think of learning the trade. He was still a farm boy and looked expectantly to his annual return to Daniel Island. Life with his grandparents had a strong maturing influence on him, and after his thirteenth birthday he made the choice which determined the rest of his professional life.

Returning from the island in the fall of 1925, Philip began to work for Peter Simmons. But his commitment was not immediate.

You know how a boy is. Thirteen years old boy then, they full of pep and that particular time I weren't too serious then. Go on the baseball diamond, play ball a while and come back over in the shop. Go cross the street and go down to the shoemaker's shop and talk to the man repairing shoes. Go down to the foot of the wharf and talk to the people who bring boats in with vegetable and shrimp and fish, seafood. And that thirteenth year is kind of a year of action for me. After the thirteenth year, then I got kind of serious in the blacksmith shop. You know six, seven, eight months after that, settle down.

Philip first learned about Peter Simmons, "the Old Man," from his reputation in the community. He was a successful black tradesman with a thriving business. A model for a young boy to follow, Peter was pointed out. In fact, he lived only a few doors down Washington Street from the house where

11

3 *Peter Simmons.*

Philip lived. His shop was located at the east end of Calhoun Street, right at the river's edge. Dawdling one day on the way home from school, Philip happened to pass by the wide entrance to Peter's smithy.

What the first thing they become interested in: the sparks they see around the shop, the blacksmith shop. And most 'prentice, most boys become apprentice on account of the action goes on around the blacksmith shop like beatin' on the red hot iron, the spark flyin' and the kids like that. I stand at the door and watch at the sparks. I said I would like to be a blacksmith some day.

Attracted by the forge, its flame and heat, the glowing red iron bars, and the cascade of sparks that rained with each hammer blow, Philip continued to hang about until he mustered the courage to ask if he could help out. To his surprise Peter agreed.

And I was going and clean the shop up and go on errands for him you know. Go out and get things and finally I would stop and spend hours in there cleaning and sweeping the shop. . . . Then the Old Man could use a boy thirteen years old at that time you know 'cause the bigger boy weren't interest. . . . He could use a boy. Hold the horse while he puttin' the shoe on. Turn the forge. He had hand forge then. We call it bellow, bellows. Turn it with hand.

Thus it was that a new apprentice was brought into the trade.

Peter Simmons (plate 3), born in St. Stephens, some forty miles north of Charleston, in 1855, spent his early childhood in slavery, grew up during Reconstruction, and earned his fortune in the Charleston district. His skills as a blacksmith and wheelwright Peter learned from his father Guy, who evidently worked as a plantation smith. Charleston farrier James Kidd suggests that Guy also had experience in the city.

Peter Simmons was the man what learned this young fellow here [Philip] the trade. Well his [Peter's] pa and my grandpa, they worked for McBride. That was years ago. The two of them was very good workers. McBride shop on the corner of Linguard and State.

Yet the earliest mention of Patrick V. McBride's shop in the *City Directory* of Charleston occurs in 1881, long after Peter had gone to work on his own. The plantation smith had to be versed in all aspects of blacksmithing, and hence it was probably under his father's direct tutelage that Peter became a general blacksmith, a craftsman capable of horseshoeing, wheelwrighting, and forging (three distinct iron trades). Very little is known of Guy Simmons except that he was a man of great strength capable of swinging two sledge hammers simultaneously, and that he was a harsh teacher. Philip retells one of Peter's stories:

He would tell the story about how his daddy used to turn the back of the hammer and hit him on the head or shoulder. . . . He [Peter] didn't hit me on the back of the hand but his daddy done it to him.

After being "clipped" a few times on the head, Peter was sure to remember the right tool to use, and he wouldn't neglect the proper steps required for specific jobs.
When Peter moved to the Charleston area

he subcontracted his services to the federal government. An ambitious government project at that time involved extensive improvements in the Charleston harbor, including the construction of large jetties to protect the entrance to the bay. A blacksmith was needed to keep the stoneworking tools in good repair.

To sit down and listen to him talk about what happened to him back then . . . they used to go to the jetty when they was building the jetty and how they worked out there. The blacksmith, they didn't had no electric welder and stuff out there but they carried the old blacksmith out there on the lighter and the anvil on the lighter and worked to the jetty there. He'd often tell me about those days. He did work for the army . . . he subworked for the army. He had enough work to keep busy.

The Charleston *City Directory* for 1890 locates Peter Simmons's shop at 37 Calhoun. Subsequent editions chart several moves along the street to the water's edge, where Philip found him. The waterfront district was thick with craftsmen, and consequently several blacksmith and wheelwright shops were clustered in that area. Within a five-block zone four Afro-American ironworkers plied their trade. When Philip made his decision to take up blacksmithing, there were several teachers to which he might have turned. There was no shortage of available role models for an ambitious black youngster in Charleston in 1925. Black craftsmen were seen just about everywhere.
It is important to understand that the success of a black man in the handskill trades was not an odd occurrence in the 1920s. Blacks had long constituted a major element in the Charleston labor force. Even before

1740 the practice of slaves hiring themselves out on their own was well established. The profits they earned were split between the slave and his owner. Thus some slaves found a means to earn money and eventually buy their freedom and that of their families. The economic outcome of this practice was that blacks, who usually accepted extremely low wages, undercut the market for white workers. From 1756 onward the protests of unemployed white craftsmen were soothed by legislation restricting the circumstances under which blacks could be trained and employed in the skilled trades, but these laws had little impact. The pattern of white patronage for black skill was so entrenched that even after Denmark Vessey's abortive revolution in 1822, which caused Charlestonians to regard all blacks with great fear and suspicion, black craftsmen were still able to earn a respectable living.

Charleston's free black class, which numbered only 586 in 1790, grew to 3,622 by 1860. In the 1848 city census this group was shown to be active in no less than fifty different occupations. The record shows further that in ironworking almost complete parity had been achieved between blacks and whites. Of 89 smiths, 45 were white, 40 were slaves, and 4 were listed as "free persons of color." Two years later some 16 free blacks listed their trade as blacksmith. Christopher Werner, the noted German ironworker who designed Charleston's famous Sword Gate, owned five slaves who carried out his commissions. Of these black craftsmen Toby Richardson is the best remembered, and he might be credited with the making of the Sword Gate and other works designed by Werner. It was not, then, at all strange that when Peter Simmons opened his shop in 1890, 9 other blacks were listed in

the same business. There were still 8 blacks who served the ironwork needs of the city some thirty-five years later when Philip began his apprenticeship. The urban tradition that he was to acquire was no less ancient than the farming and fishing traditions which he already carried. Blacksmithing had been an Afro-American craft in Charleston for almost two hundred years when Philip stood at the door of Peter's shop.

The ironwork trade also had for Philip a degree of intimacy analogous to that of farming. There were two neighborhoods in which Afro-American blacksmiths worked. F. R. Blanchard, R. M. Taylor, Fred Mechanic, and Toomer and Richardson worked on the West Side, mainly uptown. Peter Simmons, as well as David Kidd, St. Julien Logan, and Joseph Kebby, were East Siders. Since Philip knew the East Side of Charleston best, his apprenticeship appears to have been influenced by his familiarity with his neighborhood. He went to work close to home. He says, "It was convenient for me to stop in the blacksmith shop and look at him shoeing horses and stuff like that." To some degree, just as it was normal and usual for a boy raised on a farm to do farm chores, so too was it natural for Philip to go into blacksmithing. For years he had walked past some of the most elaborate and well-executed examples of the blacksmith's art. His aspirations were further developed by art class drawings. When the decision to become an ironworker was finally made, he turned to a neighbor who worked just two blocks away for guidance; he also turned to a man who was as old as his grandfather. At age seventy, Peter Simmons took on a new apprentice, who was to be his last.

In 1925 Peter Simmons had a journeyman

named McNeil working for him in addition to Philip and his own son Lonnie. Lonnie, who was a year or so older than Philip, had been working in the shop for a few years. He was glad to see a new "apprentice boy" come into the shop so he wouldn't have to take the full brunt of his father's criticism. As Lonnie recalls it:

He worked the hell out me. My mother said, "You gonna kill that boy." I say, "I'm gonna find somethin' else to do than lift these goddamn things. Find somethin' softer, lighter than lifting that, man." I was so small. I mean I had to stand up on a box to turn the forge and I had to stand up on benches to drill them wheels.

It is apparent that Peter's strictness might have been inherited from his father, Guy. He may also have inherited a conservative sense about self-reliance and the worth of hard physical labor. He seems to have avoided as long as possible the use of power tools. Lonnie comments:

My father wouldn't spend a penny for a new tool or no machine or nothin'. He made every damn thing. And man, even from the forge on up. He wouldn't buy an electric forge. Wouldn't buy electric drill. Other people, all the people on the dry dock, they had electric forge. Oh man, chisels, all his chisels, hammers, he make 'em, you know.

Philip adds further details about the work in the shop.

We drill the metal with what you call a press drill. That's against the wall with a handle. You grab holt of the handle and start turnin' it. Got a fly wheel on it and that fly wheel going around to balance it. Once you get that wheel start goin' around it's easy. We drill hole with that. We put countersink bit in it and countersink it and put the bolts in it.

The work was demanding because of the use of archaic machines and techniques and the strict discipline of the master blacksmith.

Once the two apprentices had mastered some basic techniques they were allowed to take on their own jobs, usually for the vegetable hustlers whose carts and wagons needed fixing. Any money they made from these minor jobs they were allowed to keep for themselves. Often they were paid in produce, but that was just fine, particularly if the produce was sugar cane. Since the shop was located at the foot of Calhoun, where most of the Sea Island farmers and fishermen tied up their boats before heading to the market or where they simply sat and waited for the hustlers to buy their daily supply of produce or seafood, Philip and Lonnie got to know the routines of the boatmen and vegetable sellers. Reminiscing about their apprentice days, Lonnie and Philip show the extent to which street business was their business.

Philip: That's was me and Lonnie job—fix the vegetable wagon. Get the wheel off the wheelbarrow. Put it on the side and make the wagon. People push it around here who couldn't afford horse and wagon. You know only the wealthy one could afford horse and wagon. The one that push the cart, that's is the one who we made our money [from].

Lonnie: People didn't want to take those watermelons back. Sell watermelon five or ten cents. Shrimps, they come out of a big barrel and they had a plate about this big [ten inches]. And that plate, all that stays on that plate is ten cents. They keep puttin' them on till they fall off. That would cost you at least five dollars and some cents now.

The talk of food then turns the conversation to the lighter moments in the shop.

Lonnie: We used to broil frankfurters in there. Bologna, used to broil 'em on the forge. Eat pork and beans. Put some sweet potatoes on the coals. Oh it was crude and happy.

Philip: You remember a fellow called Sunshine?

Lonnie: No, you remember Dede the drunk?

Philip: Yeah!

Lonnie: You remember Dede, he was drunk. He used to sleep in the blacksmith shop. A white fellow, he was a drunk. Papa let him. He'd sleep in there. [*Was he a watchman?*] No, he too drunk to watch.

Philip: He's an honest fellow, but he'd get drunk. He'd make his bed right there. The Old Man'd say, "Dede, you leave the place open last night?" "No, I sleep in here." He [Peter] didn't have to lock the door. Sunshine, that's a guy used to work with the Old Man. I remember when we talk about cookin'. I guess Sunshine been there about a year. He used to cook on the forge 'cause he eat too much. The restaurant didn't give him enough to eat.

Fire, flame, red-hot iron, and sparks aside, the unique cast of characters who entered the shop made it an intriguing, interesting, and at times entertaining place to work.

Peter Simmons himself was a prime attraction for people with time on their hands. Philip remembers that he used to draw crowds of bystanders.

Some of them when they ain't got nothing to do, that's what they done. Come around and hang around the shop. Yes, a lot of story went on in the blacksmith shop. Well in the blacksmith shop we can remember several of 'em. Like the guy would come by and say, "Peter, what you doin'?" Crowd around and say, "How could you work when people talkin' to you?" He said, "I don't work with my mouth, I work with my hand." And people like to shoot the breeze with him. They just love to hear him talk. He was a good storyteller.

But the work of the blacksmith shop was exciting too. What drew the largest crowds was the coordination of effort required to complete a forge weld on a large metal tire for a wagon wheel. The flat bar, approximately 2½ by ⅜ inch, was turned into a circle and the ends overlapped and heated to welding temperature. When the iron was ready, two apprentices took the role of strikers, using twelve-pound sledgehammers to hit the joint wherever they were directed by the blacksmith. Lonnie remembers it this way:

And he would strike us to death too. Striking, use two sledge hammers on welding these thing, these huge wheels. Oh, what about when he get two [strikers]. Oh, he showed off. Two men strikin' and he's in the middle. He's got a tiny hammer and he'd go boom, boom, boom, boom. But it's nothin' like hittin'. But in between he's got that little hammer just touchin' where you supposed to [hit], the next thing it's goin' to be. And he's makin' music with it, "Boom toochy-toom, toochy toom toom te toochy toochy toom." And the people just stand there. Oh, he show off, man! And at last he lay his hammer on the side like that, you stop.

James Kidd and Philip, when conversing about their youthful days as ironworkers, both admired Peter Simmons's striking of the iron.

Philip: He could play that little hammer now. Then he whistling the same time, "Toot-de-do, toot, toot." Then he take the little hammer show where to hit. Shoot. I get so weak in my hand. And when he get a real piece of iron—hold that heat a long time—I feel myself goin' down. [*James laughs.*] He say, "That's all right, stop." No, he knock the hammer and throw the hammer down. When he lay 'em down, I lay down.

James: [*Laughter*] Goddamn boy. That old man could make music can't he. Yes sir.

Philip acknowledges it was a scene much like the one described by Lonnie that drew him to the blacksmith shop. But being a witness to and a participant in striking were certainly very different. Peter tested Philip with such tasks to challenge his commitment to the trade. If after being "worked to death" he still came back, the student would likely be a good risk.

Philip progressed gradually into blacksmithing, but Lonnie dropped out after a year, turning his attention almost exclusively to his school work and musical interests. He is today a noted Chicago organist. Philip, on the other hand, dropped out of school altogether after completing only six grades.

I actually stop in the sixth grade. My father pass. I stop [school] and start workin'. My sister went to school. After the sixth grade I was promoted to the seventh grade. Working with Peter Simmons was like a trade school, we call it a trade school. But you don't know any trade school at that time. You don't know how you advancing. You don't know whether you flunking all the time or promoting yourself all the time. But I knew I weren't flunking 'cause every so often he would give me something special to do. And I feel myself that were pretty good. You could tell whether you're not doing good 'cause two of us [apprentices] in the same shop. The Old Man he'll work on the same old thing. So then I'll feel like I'm promoting myself. But you didn't had no paper work. The only paper work you had is to . . . "Boy," he'd say, "Try your hand on that. I think you can do it." By trying, I did it.

Working as closely with Peter as Philip did made the relationship between the Old Man and his apprentice very intimate. Particularly after Philip's father died, Peter must have been for Philip an authority or parent figure, if not the image of a grandfather. Philip certainly became Peter's favorite apprentice. From 1925 to 1930 several others were taken on, but none of them lasted. Other journeymen blacksmiths appeared from time to time to take up the backlog in the busy routine of Peter's shop, but none of them stayed on, even though some of them were blood-related kin like nephews and cousins. When asked if he was the favored apprentice Philip responded:

Yes, I think that's was what others didn't like about it. But I catch on things. I guess he see, some of the other 'prentice they'd want to go home just five o'clock, five-thirty. But I would stick around. Summer you get plenty of light and that what the Old Man wanted. You know, somebody'd stick around, wouldn't go.

Philip's explanation for his success is simple: he stayed, therefore he learned. Disguised in this equation is his attraction to the craft, the drudgery and long hours of physical and mental discipline required, his own skill, and the careful direction of his teacher.

Philip's first jobs were mundane chores. Today looking back, the four-year apprenticeship collapses into one event, and the dullness of the work is forgotten. The whole time is seen as a brief period of progress.

First went in the shop, the Old Man send me out [to] clean up the shop. He'd give me a few pennies, maybe a quarter. After knowing my way around, where the broom is to clean the shop and some of the tools, he got interest in me. He hire me at a dollar and a half a week. When I start getting four dollars a week, I was there three or four years—a dollar a year. It took me a dollar a year to learn. Got raises. After four years then I got a raise to four dollars.

[By] that time I'd know my way around pretty good. I could straighten iron, heat 'em and bend 'em myself. The Old Man'd step out the door and say, "Bend it, it's in the fire, hot, turn it." That's when I got eager to learn the trade, when I could do things on my own. So I would get me a little practice. So there it is. As he would get confidence, then he would launch out the bigger things or more particular thing. When I make a mistake, he would stop everything. Some people will tell

you you're wrong and keep going working. He stop everything and explain it to you.

It would be impossible to reconstruct Philip's entire educational process. It was shaped by two random factors: word-of-mouth instruction and the needs of customers. There was no logical progression from simple to complex, although Peter might have preferred such a sequence. A complicated job might have come in during the first year that was beyond Philip's comprehension. He was expected nonetheless to watch, to ask questions, to do what he could. He was allowed to experiment and to move on to more involved and complicated jobs once he had proven his competence, but there was no established scale of progress and no schedule of development to which he had to conform. His training was informal, personal, and traditional. Philip's own summary of it is most apt.

I never had a sheet of paper before me, you know, made up by my employer. He would call me when he got something particular. He wouldn't do it if I'm out of the shop. He holds it until I come. That's all. I have to catch it as it go. But for four years what do you expect? You should know your way around in four years.

If one is as motivated as Philip Simmons was, four years is enough to learn the trade of an ironworker. It is important to remember that many came to work in Peter Simmons's shop during the same time and did not leave as blacksmiths.

There is a great deal of technical knowledge relating to the metallurgical properties of iron and steel that a blacksmith must learn. Some of this knowledge pertains to the qualities of the metals, such as their strength, flexibility, fusibility, and hardness. Certain kinds of metal, like high-carbon steel, are very hard and thus good for making precision cutting tools like wood chisels, which need to hold a sharp edge. A concrete chisel must be softer in order to absorb the wear of the pavement that it cuts. A concrete chisel made of high-carbon tool steel would break in half on the first impact because that kind of metal cannot withstand high stress. The properties of various metals are best explained by scientific analysis of atomic structure and molecular behavior. Blacksmiths are, of course, not scientists, but they do have a command of basic metallurgy, at least in a practical sense. They can, indeed they must, identify the metals they work with and respond properly to their characteristics. While Philip cannot explain why cast iron is so brittle that it will shatter when struck on the anvil, he can warn a customer not to expect him to forge-weld extra decoration to a cast-iron grill. He knows that it is metallurgically impossible. One of the first abilities which one must develop in order to become a blacksmith is a skill in identifying a variety of metals and their functional properties. As Philip says, "You must know your iron. That's is the important thing." No doubt Peter Simmons introduced Philip to the mysteries of metal right from the start by having him observe that seemingly identical bars could not always be used for identical purposes.

The next basic lesson concerned the making of the fire in the hearth. Fire is the life of the forge. That it is crucial to the blacksmith is obvious at a glance. What the passerby might not notice, though, is that the blacksmith's fire is constant and controlled. The heat must be predictable. It is the iron that is worked and manipulated, not the flame. The fire should burn at a uniform, steady rate just around the area where the blower enters the hearth. Philip's directions for making a fire are clear and forthright.

Special way to make fire. It's similar to the Boy Scout. You familiar with the Boy Scout making fire? Well, it's the same thing. You must get wood, live wood. Make your fire. Then burn it down to coal, charcoal. Then when the charcoal gets red, then you put the original blacksmith coal on it and it will stay [red] like that. But you must have a pretty red coal to burn that blacksmith coal.

The phrase "original blacksmith coal" indicates that circumstances have changed significantly since Philip's apprentice days. The hard, slow-burning anthracite coal is no longer available. Indeed, almost any kind of coal has become rare in Charleston. In the old days the work was done properly with the correct materials, with the right kind of fuel.

Blacksmith coals don't have all that blast, fire. It's mostly heat with blacksmith coal. Now you could tell blacksmith coal the minute you walk in the blacksmith shop. Walk in blacksmith shop, now you'll see a little small fire but it's plenty of heat. They using a steam coals or other coals and you'll see a big blazing fire. Way up but no heat to it. See, that's the difference in a blacksmith coal. That's important too, to know your coals.

Learning the difference in fuels unlocked another mystery of blacksmithing. The chunks of coal which once were undifferentiated now became graded and labeled. Some were useful, others only wasted one's effort. Philip now had two key pieces of information vital to ironwork, the nature of metals and the nature of fuels.

Once the coal was ignited, it had to be carefully nurtured in the hearth. The coal was banked around the opening of the tuyere, the metal pipe that channels air from the blower or bellows to the fire. The coal is wet down so that it will not burn too quickly. If more fuel is needed, it is raked from the banks to the center of the hearth. If no iron requires immediate attention, the coal can be dragged back to the dampened banks and a chunk of wood set over the forge to maintain a flame. One must be careful to coordinate the air flow into the hearth with a helping of fuel. If too little fuel is added, the fire will burn out before it gets hot enough. As an apprentice Philip was often directed to pump air to the forge with a huge bellows that sat just at the edge of the hearth. It was run by a hand crank connected to a series of belts and wheels which made the bellows expand and compress. Sometimes gases would collect at the mouth of the tuyere and then be drawn into the lungs of the bellows, where the gases might explode with the force of an automobile backfire. If that happened the fire could be blown out; certainly the valves in the bellows would have to be adjusted. The main nuisance of these explosions was the loss of heat. Philip remembered that they usually occurred at the most inopportune moments.

Here you got the weld ready. The Old Man say, "Get that end in ther boy, get that end in." When I walk, the bellow pop. Now you just want about fifteen seconds for 'em [the heat] to be perfect.

Fans with a constant blowing action that eliminated the backfire problem were available in 1925, but Peter preferred his old-time machinery. Philip, however, never was happy with the labor of turning the bellows crank. When he became a full-fledged smith, one of his first acts was to install an electric forge blower.

Once he mastered his way about the shop and could tend the fire, it was time for Philip to learn the names and uses of the various tools. Some were obvious, like the hammers that were used to shape the iron and the anvil that was the striking surface. But there were hundreds more. Many were highly specialized and rarely used. Still he had to know their purpose. One of the best ways to learn about the tools was to put them in order, put them in their place in the shop.

I had a lot of responsibility. . . . Puttin' things away, puttin' tools in the proper place. I was handy when needed. I had to hand him different tools and hold different things. Old Man say, "Where's my punch?" I get it.

Many of the tools were classed as anvil tools. These were an assortment of devices which were set into slots in the top of the anvil.

You got about maybe thirty or forty pieces of different swaging tool. It changes according to the shape of the iron that we making. That's my job. When he pull the hot iron out, I must have the right swage there. Hardys and stuff like that, tools that go in the anvil. You see, anvil got from two to three holes in it and those are the holes to hold the hardys and swages, bottom and top swages. But you sit it right in that hole. It all fit in that hole, one hole. You can have so many different sides. You can have a chisel side, a pointed side, a round side, a square side, and all those different shape. . . . You put the swage on there, some swage will make the iron round, some you make square, some you make it flat, but you got a swage for different kinds of work, top and bottom. The bottom one goes down in the anvil and the top one you hold it in your hand like a hammer. All of 'em got a handle like a hammer handle.

And so, on and on Philip learned the variety of tools: punches, chisels, tongs, drifts, sets, fullers. At the same time he gained a sense of personal tradition, for many of Peter's tools had belonged to his father.

He [Peter] loved good tool and several tool he would say, "Keep this tool" and "because it's my father's." So we said, "We'll keep it for a souvenir." He said, "No, just use it." We using it. I got several pieces back there [in the shop] now I'm using. So see that tool could be about a hundred and some-odd years old.

Peter was concerned that his tools be cared for properly, not only for the longevity of the tool, but for the safety of his apprentice. Working with burning metal somewhere near 2,300° F., striking it with violent blows, created a certain degree of danger from sparks and chips of iron. Even when cutting cold iron Peter cautioned Philip to "muffle it down," that is, to hold the chisel with an open palm and close to the bar. This way, if the short end of the iron were to fly up with the impact of the hammer and chisel, the fragment would not get away and strike the apprentice in the eye or about the face. It was important, indeed crucial, to work safely.

Well the first thing he would do, he would teach safety. Tell you how to use it [the tool] the safe way. How to protect yourself. And every tool that we use in the shop at my early age as a blacksmith, even before I used a tool, he would tell me how to handle it. There's several ways to handle tools, but you would get hurt with it, 'specially in a blacksmith's shop. First would teach safety, then he would go ahead and tell me, teach me the advantage of different tools. You could take one tools and use it, it would be to no advantage. . . . Every piece of work there's a special tool and he don't want you to use a hollow tool if you making your work with flat iron. Every tool to every work. That's what's important. I look for a long time. I stand around for over four or five months. Just stand around looking before I touch a piece of tool. I was aware of all that tool before I start using them.

Because Peter Simmons worked in all branches of smithery, he had a broad range of customers. His shop location encouraged this variety. He was accessible to city dwellers and Sea Islanders alike, repairing wagons for urbanites and plows for farmers. With his shop situated at the river's edge, he received a number of requests for boat iron. The routine of the docks, which involved loading ships and river flats, also provided him with work. He fabricated the iron frames with sling hooks called "drays," which were placed under the bundles of cargo to make it easier to hoist them and place them in the hold. Through the doors of the shop came people from all social backgrounds and almost all professions. In learning to be a blacksmith, Philip also became familiar with the full range of business activity in the Charleston area.

Learned to make horseshoes and wagons. Well in those days the blacksmith had to do woodwork too. You know, Peter Simmons, he done the woodwork himself and the ironwork. First you build the wagon, then you iron it off. You see some wagon got about as much iron on it as what wood. You had to know a lot about carpentry as well as blacksmithing because in those days they couldn't afford to furnish a blacksmith and a carpenter too. You had to do both. We had two different type of wagon. We had a Tennessee wagon and we had a city wagon. The country wagon [Tennessee type] has a different turntable on it. You can turn more shorter at the corner.

We makes plows. He bought some it and he made some it. Sometime a farmer will come to you with something they arrive from their own ideas, you know. You had some smart farmers. They know what they wanted. Sometimes they buy the sweep turn plow. And they just do it one way, they don't do it the other way. They bring that plow to you and we changes it. A real farmer, doing it for a living, doing it any length of time, he knows his plow. He knows what he want. They come to you an' you got to come up with the same thing.

Now even boats, we used to make a lot of iron for boats, to iron out boats. Make the mast, the stair, the ring hooks, and all of that. We made a lot of anchors, rudders, steel keels, and all like that for boats. We made it for boats as well as wagons now. We done as much iron on boats as the wagon. You'd either go down to the wharf and make the drawing or they bring it up to you. Some of the boatsmen there know what they want. No if an' and about it. They couldn't make it, but they knows what they want.

Some of the other important work I have done is to make timber carts. They didn't have tractors then to snake the log out of the woods then. You had to have timber cart and big horse to pull it. We have to keep that up. We make the tines [to grip the log]. We make the timber cart and all. Two wheels and one shaft. We pulled by two horses. Tongue came out of the shaft. That was some of the big things we done back then.

I shoed several horses. I work as apprentice. Prepare the hoof for the horseshoe. Then as I develop, I start help shoeing. I shoe horse about one year maybe. That's all. It was all pretty well gone. Different shoes, no two horses put the same. Heat 'em up, change 'em, bend 'em. Even if you buy the shoe from the factory or the merchant you have to set all the shoes.

Wainwright, forger, boatwright, cartwright, farrier—all were trades in the domain of the smith, and Philip learned them all.

Since Charleston did not have an extensive system of smoothly paved roads until after 1925, wagons were constantly brought in for repair. The uneven cobblestones jolted and wracked the wagon bodies; spokes and hubs were jarred and splintered. Consequently, many Charleston smiths advertised themselves as "Blacksmiths and Wheelwrights." Peter Simmons was no exception. Indeed, his most spectacular display of talent was, as we have seen, making metal tires for wagon wheels. Hence Philip became particularly proficient as a wheelwright.

Tire got to be smaller than the wheel. The wood part is the wheel, the tire is the iron rim. Just like you would put a tire on automobile; the rim is metal, the tire is rubber. Opposite with wagon wheel. Rim is the wood and the tire is the metal. You make it smaller than the wheel. If your tire is three-eighths inch [thick], you make it three-eighths inch smaller. If your tire is one-half inch, you make it one-half inch smaller. If the tire is one-quarter inch, you make it one-quarter inch smaller. And you heat it [the tire]. When you heat it, it expand. Heat it round and round so it can expand. We put it [on the] outside. Put fire on it. Put paper down, just like Boy Scout make a fire. Get it hot. Sometime you get it red. Sometimes you get it too hot, got to let it cool off some before you put it on. All metal expands and when you cool 'em off he shrinks back up. Set it right on the rim and pull it on. Got to pull it you know, then you put it in the water. Taken a can of water, make a little trench, and turn the wheel around in the water.

Occasionally the tire will become too tight on the wheel. The wooden parts of a wheel, if not properly seasoned, may expand once the tire is set. As they expand the spokes split or the rim cracks and warps. In situations like this the tire must be stretched. After it is removed from the wheel, a section of the tire is heated until red and then beaten with a swaging tool on the inner surface. The tire then has a corrugated appearance, but it is also slightly larger in diameter and should fit snugly without too much pressure. Whether shrinking tires or stretching them, Philip learned that the blacksmith is the master of metal. It takes the shape that he commands. It may be difficult to beat into shape, but the blacksmith is sure that the iron will submit to his design.

Four years passed with Philip growing not only in physical stature, but also growing in Peter's esteem. The skinny shop boy had become a helpful and sure-handed apprentice.

He went to his assigned tasks with enthusiasm and took instruction well. He seemed to have natural talent. Philip says that he loved blacksmithing. Other apprentices came into the shop but didn't finish out their tenure as trainees. Philip stayed on,

because I fall in love with it. A little more action in that than other work. Money wasn't important. Something about the blacksmith I had love. Sparks and other things. Coolin' down the hot iron. Something with excitement. Something that would get you excited in the blacksmith would get you excited nowhere else.

His excitement led him to be successful in learning the trade. His success, in turn, was rewarded with responsibility, with new assignments that were more complicated and difficult, but more fascinating and challenging. Philip earned these jobs because he displayed initiative. He brought small repair jobs to the shop on his own, like broken charcoal irons that required new handle brackets. Acts like this signaled to Peter that Philip's apprenticeship was nearing its end. The proof that Philip had matured into a journeyman blacksmith came in 1931.

I was gettin' about four dollars a week after I become knowin' my way around and could take care of the shop by myself when the Old Man go away. Okay. He raise me to four dollars. He see my ability and he know me. He knew everything about my ability then. So then I had a chance to do things in the shop that he would let the 'prentice do and here I feel my ability.

He took sick and went into the hospital. I was about nineteen years old then. I were getting four dollars a week. He went into the hospital and he stayed in there, the hospital, and stayed home. Altogether was two months. And he stayed in the hospital two months and I run the shop. Then I knew I'm goin' to be a blacksmith. Until then . . . I just knew I goin' be a blacksmith after I run the shop myself and he was satisfied.

The first job I done, I fixed some tubs. You know the schooner, you know what a schooner is? Well, they used to bring coals to the Johnson Coal Yard, Pochahontas coal down there to the waterfront, right there by the fish dock. Johnson Coal Yard was right in that spot and big schooner with sail would come in there. And they had buckets, big buckets hold about a ton and a half. And it's metal bucket and beatin' it against the wharf and they get holes in it. So I had to repair them. And they sent me, from the coal yard, they sent me three tubs. And I repair those three tubs and sent 'em a bill for thirty-five dollars. Three metal tubs repaired and they send for it with the truck and took it there. And I carried the bill. I made the bill myself. That's why I'm tellin' you now I feel myself then as a blacksmith. I'm going be a blacksmith.

And they give me and they wrote the check for thirty-five dollars and I came back. Peter Simmons was in the hospital. I took the check there and he signed it for thirty-five dollars. And I took it and cash it and one-half of thirty-five is what? Seventeen dollars and a half and he gave me the seventeen-fifty and kept seventeen-fifty. Then I knew that I was goin' make that my career.

From four dollars to seventeen telling me that that's what you want to. That's was the most money I ever had in my life. In my life and I was nineteen years old. In nineteen years I earned seventeen dollars. That's was in Roosevelt time. Hoover time or Roosevelt? In the Depression. You from four dollars. Here you got a raise to seventeen dollars. That's seventeen on that one job. It took me a week to do it though. It took me a good week to do it.

Those was big buckets, look like a metal tub. We didn't weld it to patch it. We patch it but we didn't weld it. We put rivets in it. We cut the patch out and put hole in it and rivet it together. Was a little harder than welding. Now you can do it three times faster now than you done it at that time.

But the thing what this say to me, "This what you should be. Money's in it." Seventeen dollars a week. That's what really happened. And [in] the whole blacksmith trade and the whole time I was in there that's was the most amazing thing to me—that particular day.

Peter Simmons returned to work to find Philip a fully trained and tested blacksmith. For four more years Peter maintained control of the shop at 4 Calhoun, but with a full partner. He was still the boss, and as Philip notes, "he called the ball and strike." But in the early 1930s, Peter moved across the river to the town of Mt. Pleasant. There he continued to run a small blacksmith shop. Peter's partner was eventually to be the blacksmith of Charleston. At the age of twenty-three Philip was about to embark on an illustrious career as a craftsman, a noble future for a former Sea Island farm boy.

3

"Then I Went into It Forcibly"

PHILIP LEARNED his trade from every smith who had something to teach him. While he was apprenticed to Peter Simmons he also managed to visit the other blacksmiths in Charleston. "Ducky" Logan, for instance, had a shop just across the street from Peter, and Philip would on occasion go over and lend a hand, particularly with a heavy assignment like striking wagon wheels. He also visited David Kidd, a local farrier.

I went and help out Kidd, David Kidd down on Guignard Street. David Kidd's shop was down by the big market. Ice house and all in there at that particular time. After he passed I used to go help out John Morant. He took the blacksmith over after the death of David Kidd. David Kidd, when I was a boy and I were interested, he always thought he could help me. And he did.

Philip dropped in at all the blacksmith shops from time to time. He not only pitched in when asked, burnishing his reputation as a forthright and dependable worker, but he also learned more about the trade. He saw different techniques for handling common jobs as well as different jobs which never came to Peter's shop. The tools used were also different, as each smith had a slightly personalized approach and special techniques requiring tongs with angled jaws, a different size punch, and so on. During the late 1920s Philip became friendly with almost all the ironworkers of Charleston, white or black. For example, he got to know the shop of the German blacksmith Frederic Ortmann.

He was on Meeting Street when he pass. Blacksmiths then they always had time to visit. Not all of 'em, 'cause all of 'em wasn't on good terms. I was on good terms with all the blacksmiths. After I became a man, twenty-one and on, I was on good terms with all of 'em. I could go in the blacksmith shop and borrow anything, any kind of tool I wanted. They'd come to me. I had a good relationship with them that didn't exist among all of 'em.

This trusting friendship within his profession was the beginning of Philip's city-wide fame as an honest and dependable craftsman. Once he became a regular visitor, he was regarded as a shop fixture. The other smiths sometimes paid no attention to him at all and just went about their work. But all the time Philip was learning the secrets of the trade.

I never asked questions. I'd go 'round and observe. "Well," I say, "that's that," once I see how they done it. I always were observant and some of the blacksmith then didn't mind. I pick up things, several things. So that's the way you learn.

Before Peter turned the shop over to Philip, he made him a full partner who earned the same wage as he earned. They were equals in the shop, although Peter's seniority gave him the edge. In 1935, however, Peter thought it best to work at a slower pace, so he moved to Mt. Pleasant (see map facing p. 1).

He move over there at the age of eighty. So you see he died at the age of ninety-eight. So he was over there about eighteen or nineteen years. I stayed here in Charleston. I run it [the shop] as part owner and later on then I just take it over after he got old, after he got to an age that he couldn't speed cut, that he couldn't do as he want to do.

But Peter's leaving did not sever the bonds between the two men. Ferry boats regularly crossed the six miles of bay to Mt. Pleasant, and Philip made regular visits.

I went over in the evening and help him. I work over here and at four o'clock I'd catch the ferry boat and go over there and help him. He was an ambitious person. He wanted to be independent. He didn't want to see nobody do anything for him. So when I'd go over there I just do it. I don't even ask him.

Philip would step off the boat and walk the few blocks to Peter's new quarters. A smithy was set up in a shed in the back yard near the intersection of two alleys. After firing up the forge, Philip would set to work on whatever seemed to require repair or service. Peter sometimes appeared a little grumpy that his old helper was taking over his new preserve, but inwardly he appreciated Philip's concern. Perhaps he even marveled that after working the whole day, Philip would take what amounted to a second job. Philip, however, saw his actions differently.

I couldn't call it exactly two full-time job 'cause I'd go over in the evening, maybe four, four-thirty and I work until about seven or eight o'clock. But I done that so he wouldn't work too hard. Like the sun is now [in the summer], I can go over there at four o'clock and give him a good three, three and a half hours. Sometime I stay and sometime I come back. I just thought he'd worth that for what he had done for me.

The bond between the master and apprentice was not broken, even after the apprentice passed the crucial tests along the path to maturity. Philip continued to acknowledge that his competence as a craftsman was due to the excellence of his teacher, and he willingly and generously offered his labor in order to reduce Peter's burden. At age twenty-three, Philip was full of his future. It was sure to be challenging and difficult, but he would face it with confidence, the greatest gift that Peter Simmons had been able to give.

Had it not been for his large store of self-confidence, Philip, like so many others, might well have abandoned the blacksmith trade. There was a marked decline in the number of ironworkers during the period in which he assumed full control of the shop. Most of the blacksmith shops closed down and the smiths sought work elsewhere. Some retired; others moved north. Their apprentices turned mainly to automobile repairs. While industrialism had a late impact on traditional aspects of Charleston's business scene (horse-drawn wagons were still common on the docks in the thirties), the decline in demand for the skills of the handworker did eventually come nonetheless. Philip had received his first inkling of the precarious position of the old-timey trade while still an apprentice. One of Peter's best customers, Willie Rooney, began to replace his horses and wagons with trucks.

He have horses, drays . . . he got a secondhand truck first. Taken the place of a wagon. He bought a secondhand White. A year later he got a new . . . he bought a new Ford truck. So the driver say, "What you gonna do, ain't have no wagon no more, all the wagon gone? What you gonna do, what you gonna do?" I wasn't no more than about sixteen years old. I feel kind of skeptical after people coming around talking about the wagons are going out. No wagon. You know what that mean, no horse. No wheels to repair or built. Then what I goin' do? 'Cause those four trade I tied to it, general blacksmith. [I can] make chisel, that come once a while you know. So I didn't have sense to look beyond that chisel. And I think about these power tools you know. Ain't have to make no more chisels or stuff like that. But I couldn't see beyond the wagon. What goin' to take the place, what you can launch into after the wagons and all. Once they get off the street, I couldn't see it.

His fears were calmed by Peter's advice: "The Old Man say, 'Boy don't worry.' He say, 'There'll always be work here for a blacksmith. Don't worry about it.' " Philip did not press for a more detailed explanation. He waited and watched and saw that sure enough, the Old Man was right. The work continued to come in. The doomsayers were proven wrong.

The first trucks were very similar to wagons in appearance. They were essentially truck bodies set on a wagon frame, but without the hitch for a team. Willie Rooney's driver, who had questioned Peter about the future of blacksmithing, was no doubt surprised when a week after the first truck was purchased, it was brought into the blacksmith shop to be "ironed off." Philip remembers it well.

Fixed the body of it. Put stakes on it. Make pocket for stakes on the outside. Put the iron pockets on the side when you goin' to stick the stakes in just like a wagon. Same principle, same identical thing.

The trade thus continued with barely an interruption. A major shift in technology from the horse-drawn to the motor-driven had occurred without the slightest variation in the work of the blacksmith. It was, however, an important moment for blacksmithing, since it provided yet another area in which to work if one could but seize the opportunity. Following Peter's example, Philip took full advantage of the chance to work on truck and car bodies and thus found that his old-fashioned skills could be put to contemporary use. Peter's advice had proven to be an accurate prophecy—there always was work for the blacksmith.

Because Philip was already a blacksmith when the automotive age emerged in Charleston, any work done on trucks and cars registers in his mind as blacksmithing. Much of what he had done in his career would be classed by some as "body work," but he sees it as related to his ironworking background.

It's a funny thing about the blacksmith. You know the blacksmith trade give you so many idea about other trade in iron now like the ornamental ironworking or the angle smith or the automobile shop area, the automobile builder, and building truck bodies. You see the blacksmith shop is the father of all those trades. I didn't knew it until I start working there and I see people after they finish blacksmithing go into the automobile shop and build trailers and stuff like that, you know. So the blacksmith shop is the father of all other trade.

New kinds of work with cars and trucks thus seem to Philip to be a natural extension of his chosen trade. While he abandoned the wheelwrighting and carriage-building fields of his profession, he maintained, in his mind, his identity as a blacksmith.

Look at some of the automobile, the trailer hitch we put on the back of automobile. And we still repair trucks right now. The factory turn it out but after it get here, broke, then. You see that's where you get [a] hand on the blacksmith quicker. 'Specially the area where you can find a blacksmith. They don't pass the blacksmith by to go to the truck builder. When the blacksmith make [something], he can think back a hundred years.

The work of a blacksmith will always be marked by antiquity regardless of the task. The process of shaping iron is basic to the blacksmith's identity. Regardless of the circumstance, if Philip drills, welds, bends, punches, twists, cuts, or forges a piece of metal, he is following his trade. Horseshoe, wagon wheel, or trailer hitch, it is all the work of a blacksmith. Equipped with this imperialistic attitude—that all iron and steel is circumscribed by the boundaries of his trade—Philip was able to move into the

contemporary age with little difficulty. He faced the changing times of the 1930s and 1940s with confidence and added new options to his repertoire of jobs. While the business of other blacksmiths dwindled to nothing, Philip's work expanded and flourished. His success in an archaic profession can be credited in some degree to the encouragement given him by Peter Simmons, who trained him to master iron so well that mere changes in technology would present only a slight problem of adjustment.

Enough standard blacksmithing jobs still came Philip's way while he worked on cars and trucks so that his traditional skills did not wither from lack of use. Many of the tradesmen along the waterfront needed special tools which a blacksmith alone could furnish.

Every so often you see somebody bring some chisels in there and throw it down and say he wants 'em sharpened, or bring the raw tool iron and throw it down and say he want some chisel, want some special chisels to be made. Brick layers and shipwright, they still use caulking tool. I was working for about four of the longshoremen's stevedoring company. I used to make all the bars, drills, and stuff. We made hooks. At that time we got make all your shark hooks. You could buy 'em but they would straighten out. It weren't what they want, so. . . . There were cotton hooks and shark hooks. They were tool steel. They would break before they'd bend. You know what you gig fish with? We made all of those. That time there a big demand for those things then.

Just as Philip repaired and maintained wagons and truck and car bodies, he also repaired the tools he made. In the spring both city dwellers and farmers brought their hoes to the shop. They had been broken during the harvest and had lain around all winter. Now with the coming of planting season, the hoes needed to be rehafted and fitted with new handles, or at least sharpened. Chisels used with "automatic air drillers on the street" are in constant need of repointing.

The old one, when you goin' to sharpen you put it in the fire. Finish drawin' it out, the part that you've actually worn off in working. So repairing chisel is not like making it. Making 'em you've got to start from scratch. You got the whole butt end there to beat down. But repairing them you got a part point there.

Other tools Philip made and repaired commonly were plumber's caulking chisels, stoking irons and shovels for colliers, and mason's trowels. Toolmaking had once filled most of Philip's schedule, but in the late thirties and early forties such demands became more intermittent. Because they did not stop altogether, Philip could still perform as an old-fashioned blacksmith often enough to keep his sense of tradition alive.

Philip had to move his shop location several times just as he began to emerge as an independent craftsman. The location at 4 Calhoun Street was taken over by the city and the building leveled to make way for a park that would serve a recently constructed housing project. Philip shifted his business to a new site across the street and a few doors down to number 9, but stayed there only a year and a few months. When he found a spot on Alexander Street just a few doors up from Calhoun, he stayed put for fifteen years. This location was outside of what could be called

the industrial or waterfront area. Here his business was nestled in a neighborhood of Charleston single houses.

The new domestic setting may have had an important influence, for it was on Alexander Street that Philip began to specialize in decorative blacksmithing. He still did general forging, occasionally made tools, and even repaired broken plows and wagons. His waterfront clients could still find him, but the times had changed enough that ordinary hardware was usually purchased at a store rather than ordered from a blacksmith. Wagons, as we have seen, were on the way out, and automotive specialists were becoming more common.

Only an esoteric area within the blacksmith's domain, ornamental wrought iron, remained unchallenged. Looking out from his shop at a row of Charleston houses—some plain and some decorated—Philip realized that decorative ironwork was what his future depended upon. His current enthusiasm is a vindication of his early assessment.

There's a big demand for iron gates, big demand now. There's always gonna be for that type of work: iron gates, window grills, porch railings, and columns. As long as they gonna build a house, they gonna put iron on it in Charleston. 'Specially in Charleston.

Although he had not done very much ornamental work, he knew that he had the skill to do it. All that remained was to develop a new clientele. Philip summarizes this important shift in his career succinctly.

He [Peter Simmons] didn't do too much ornamental work, but after I took over then I

4   *Gate, corner of
Ashe and Spring
streets.*

went into it forcibly. Well, I found myself into it all the way until today.

Philip's entry into the decorative field of ironwork was gradual—at least that is how he remembers it.

And after the horse-buggy wheels go, something come to me. Say, "You a blacksmith, you can bend those iron just like most pattern. You could do a better job. You could make it look like the old originals 'cause you got your forge." When I got on the forge and people come in. "Mr. Simmons, I got a broken place in my gate. Could you repair it?" I go there and repair it. Didn't have no electric drill. You take the hand drill and rivet. Had [to] rivet them scrolls in. It was hard riveting them. And finally I work and I keep my pride, do my work with pride. Finally I start with one piece, repairing gate. Then start making the parts to put in 'em, like the man got down to the Sword Gate. Finally, I start making the whole gate. I said, "I got the forge and I turn 'em out like the old original." Finally people going for that stuff. "Go ahead make it like that. That's what we want."

To move from repairing gates to making the whole gate took less than a year. Peter Simmons had done repairs of decorative wrought-iron gates and fences, as well as making fancy lamp brackets, so Philip had some basic knowledge of the required fabrication processes. Yet in all his years Peter had never ventured fully into ornamental ironwork; he didn't have to. Philip, in contrast, plunged boldly into this specialty, not only as a way to make the blacksmithing trade support him, but as a way of taking charge of his shop. He had been on his own for a few years, but in a role and place that bore the stamp of his predecessor. In this Alexander Street shop, Philip made his first outright claim to leadership in blacksmithing.

I was the boss then. That's the most complicated thing. But it wasn't long before it come natural to me. It's just a matter of working with people right on.

Philip does not have elaborate recollections about his long career in decorative ironwork, a career which began in 1939 and continues even now. This might at first seem odd, but when we consider that he is a man of actions and deeds, not a man of words, we soon recognize that the record of his career is seen in the numerous examples of decorative wrought iron which he designed and forged. Charleston and the surrounding area is an archive of Philip's progress and achievement. He estimates that in the forty years since he first took a decorative commission, he has turned out more than two hundred gates. He has also made balconies, stair rails, window grills, and fences. Much of his work has been for what he calls "poor folks" who have only wanted a simple walkway gate to keep out stray dogs. Such examples of his craft are plain and simple with few decorative scrolls and, for Philip, are easily dismissed. When passing one of these gates he is likely to recall, "That's one of my jobs, but it's not important" (plate 4). He does classify certain works as important, calling them "fancy." Philip chose the ironwork in the following portfolio. The arrangement is chronological so that the pieces may be understood as historical statements as well as works of art.

A Portfolio
of Decorative Ironwork
by Philip Simmons

Plates 5, 6

9 STOLLS ALLEY

In 1938 a client from Georgetown, a coastal
city some sixty miles north of Charleston,
commissioned Philip to design and fabricate
an iron stair rail for the interior of his house.
This railing was Philip's first decorative piece,
and it eventually led to other requests for
custom-made ornamental work. His local
reputation had rested mainly on his expertise
as a repairman and maker of tools, but after
making the gate at 9 Stolls Alley, he was
known for decorative iron. This walkway gate
served as a kind of billboard for his shop.
Even now it draws attention, and since the
house to which it is attached is one of the few
examples of pre–Revolutionary War
architecture still standing, the gate is assumed
to be of a similar vintage. This error is
excusable inasmuch as the techniques of
construction are easily as old as the
eighteenth century, and the design closely
follows nineteenth-century modes. It is most
significant that from the very beginning Philip
was able to control the forms and motifs of the
Charleston tradition. His first fancy gate is a
distinguished example of the local style; his
mastery of the tradition is so complete that his
work is often mistaken for the work of the
past centuries. Philip says of Charleston
ironwork:

The old work was good. The scrolls were
curved nice and round. If you see it curve like
that it's either two hundred years old or I
done it.

5   *Gate, 9 Stolls Alley.*

6   *Detail of gate, 9 Stolls Alley.*

This gate is composed of two sections, an overthrow of pointed bars and the screen of the gate itself. The overthrow is fan-shaped with a **C** scroll at its center set in a small semicircle. From this iron arc radiate ten round bars which penetrate a larger arc and terminate in flat triangular spear points. Small **C** scrolls are attached to the inside of the larger arc between the bars. The screen of the gate is composed of **S** and **C** curves. Paired **S** scrolls that form a lyre are joined by an arching piece of iron which simultaneously takes the shape of a **C** scroll and blends into the **S**'s. The sides of the screen are accented by four **J** curves. The spear motif seen in the overthrow is repeated in the screen at the ends of a square bar which runs vertically through the center of the gate. Three spear points are also riveted to the top edge of the screen. The decorative scrolls are not attached directly to the frame of the gate, but are held away from it by iron balls. The entire composition thus might appear to float within its borders.

Plate 7

65 ALEXANDER STREET

A woman who lived across the street from Philip's shop learned that he did decorative work. She came over and asked if he could make her "one of them pretty gates like they got downtown." Philip replied that he could,

7   *Gate, 65 Alexander Street.*

and he made what he calls a "harp gate." It has a lyre form with three pointed bars for strings as its main decorative element. This short walkway gate is topped by a pair of large horizontal **S** curves with a fleur-de-lis mounted between them. The lyre panel is formed by reversed **S** scrolls spread apart by a flat **C** scroll and is accented by **J** curves on each side. The round bars that form the lyre strings are drawn down into leaflike spear points at the top, while the bottoms are welded to the **S** scroll. The base panel of the gate is a horizontal rectangle filled with a complex **S** curve design. Two scrolls appear to overlap. The design is, in fact, a series of **J** curves manipulated to look like interconnected **S**'s.

Plates 8, 9

2 STOLLS ALLEY

This riveted gate obviously predates Philip's acquisition of an electric welder, and the tiny lumps of the flattened metal visible at every attachment of a decorative element are considered a mark of the old-fashioned style from the past century. The gate screen is divided into quadrants with the same ornamental motif recurring in each section. **S** and **J** curves of several sizes are intricately combined. At the outer edge, two **S** scrolls are paired and matched against a large **S** which abuts the central axis. The spaces within and around this scroll are filled by pairs of long and short **J**'s which are beat into reverse curves (a shape rarely used). The configuration

8   *Detail of gate, 2 Stolls Alley*

struck Philip as strange, but he submitted to the demands of his client.

Some peoples there got dogs and you got to fill those thing [gate sections] up. They didn't want bars. Said, "Don't want the bars. I want some scrolls down there. The whole gate, I want scroll. But I don't want the cats and dogs to come in." Have to fill the space and that's why it look so much different. But if I were going to make it for myself, I wouldn't make it that way. I wouldn't try to keep the dog and cats out.

The four decorative panels are arranged so that the motifs are symmetrical both horizontally and vertically. Careful inspection of the gate might reveal that the overall design pattern within the complicated arrangement of scrolls could be interpreted as two lyre shapes with surrounding decorative curves. This was the basic design of Philip's previous gates. The top edge of the gate has a paired arrangement of **J** and **S** curves with a wavy fleur-de-lis in between them.

9   *Gate, 2 Stolls Alley.*

Plates 10–12

5 STOLLS ALLEY

The quality of Philip's two previous gates on Stolls Alley led to another commission on the alley for a gate and an elaborate porch railing. The walkway gate is crowned by a pair of **S** curves embracing a wavy fleur-de-lis. Its screen is filled with a floral composition which runs from the base of the gate and terminates with outlined tulips in the top corners. The stalks of the flowers are composed of loops, **J** curves, and bent slats which terminate with iron balls. This imaginative profusion of ironwork leaves fills in the two sides of the screen and disguises the central motif that forms the basis for the gate's design. At the center are two large **S** curves which overlap each other. Philip remembers that

this [gate] is two halves. What we had in mind was to fill it. And after we make the large one [the **S** curves] we have to fill it in with small scroll. Fill in the sides.

Thus for him the attractive leaf composition which holds the viewer's attention is of secondary importance. It is a fill unit, used between the large central curves to fill up the empty space.

The porch railing, now partially obscured by a leafy bush, is made up of five sections: a short ascending stair rail, an elaborate panel with flanking motifs, and a terminating unit that fences in the end of a brick landing. A leafy scheme was rendered in the central section, which employed the same design strategy used on the gate that is a mere six feet away. This panel is approximately

10   Gate, 5 Stolls Alley.

12   *Porch railing, 5 Stolls Alley.*

rectangular with a slightly curved top and is filled with **J** curves, loops, and bent slats. In the middle, pairs of scrolls of increasing length are stacked four high so that they appear to be a series of inclusive **C**'s. A leaf shape outlined with slats of flat iron stock fills in the space between the curl of the volute and the unbent end of the scroll. To either side of this assemblage of leaves is a series of sweeping **J** scrolls, each mated with a leaf set at the end of a long stem. In between the edge and central decoration are sets of loop forms with iron balls attached to them. The flanking panels, although ornamented with scroll designs based on the lyre pattern, are quite plain by comparison. They are interesting nevertheless, because these decorative units are aligned horizontally in an asymmetrical position and hence break an important local convention that demands a balanced design.

11   *Detail of gate, 5 Stolls Alley.*

Plate 13

125 1/2 TRADD STREET

The walkway gate at 125½ Tradd Street is
rather tall for a garden entry. Perhaps the
height of the extant brick pillars on either
side was an important determinant. Philip
noted that the floral motifs used were similar
to those he had used in the gate at 5 Stolls
Alley. In them we find a combination of loops
and **J** curves. They surround an octagonal
design which was rescued from an abandoned
grill. The gate's design was a simple matter of
securing the grill in the center of the screen
and placing the leaf compositions above and
below it; in Philip's words, "The one on Tradd
Street definitely was a grill. Was grill first and
we cut it up."

Seen from the front, the upper leaf element
runs from right to left, while the lower one
begins at the left edge and ends on the right
side. The asymmetry of this composition is
disguised by the round shape of the center
grill. The leaf elements seem to swirl around
it in a counterclockwise direction. The top of
the gate is decorated with **S**'s and a loop form
that repeats the shape of some of the leaf
units found in the screen. Cast-offs and
creations are joined in this gate, a reminder
that salvage and service are two primary
modes of the blacksmith's trade.

13   Gate, 125½ Tradd Street.

Plates 14, 15

8 LAMBOLL STREET

This combination of drive and walkway gates again features the loop-and-scroll configuration of leafy stems. The driveway section (plate 14) appears to be designed as if it were two smaller gates, since it is topped by two independent sets of **S** curves. Usually the decoration on one half of a gate is the opposite of its mate, so that when the gate is closed a single design appears. The screen is sectioned into quadrants, with the same decorative cluster employed in each quarter. Since there are no dividing bars to mark the different sections, the design seems more detailed and complicated. The pattern is not very different from the one used at 125½ Tradd Street and 5 Stolls Alley. A stalk with leaflike scrolls was constructed with loops, **J** curves, and **C** scrolls.

The decoration in the walkway gate (plate 15) is arranged in the same general pattern as the driveway gate, but with notable differences. The loop has been omitted and a **C** curve with its ends bent to touch has been substituted for the **J** scrolls. The loop motif that is seen along the outer edge of the other gate has also been left out. Because the garden gate is shorter, the upper quadrants have been compressed, and a unit of only three scrolls is used to fill in the space. The same **C** and **J** curves found elsewhere in the driveway gate are employed, but obviously without the same elaboration. In the screen of the railing to which the walkway gate is attached, the scroll elements are set in yet another configuration. The patterns used in the bottom of the garden gate were deployed in the top part of the railing, but aligned so that the leaf stalk points downward. The bottom section then features units of **J** curves embracing a loop (the first third of the upper unit in the garden gate). This lower ornamental motif points upward so that the looping finials touch.

The variation possible within the limitations of three basic motifs is most clearly shown in this piece. The reconfiguring of those motifs provides a basis for unique creativity, so that each gate section is different while the repetition of decorative elements gives unity to the entire composition.

14   Driveway gate, 8 Lamboll Street.

15   Garden gate, 8 Lamboll Street.

Plates 16, 17

36 MEETING STREET

The decorative panels in this combination drive and walkway gate have a marked feeling of antiquity. They are, in fact, elements from a much older gate or railing that Philip rescued from the junk pile and placed in this contemporary configuration. These attractive assemblages of water leaves with crimped edges and scrolls with ends that curl into two complete turns closely resemble a 1795 window grill from the old Bank of South Carolina. They might even be that old. Certainly these ornamental units are very different from the rest of the gate's screens, which consist solely of unadorned straight bars. Philip has placed a band of crisscrossed bars just below each decorative panel to serve as a counterpoint to the curvaceous leaves and scrolls.

It is interesting to note that even while

16   Gate, 36 Meeting Street.

17   *Detail of gate, 36 Meeting Street.*

Philip's work as a blacksmith preserves a local craft tradition, he will also take pains to preserve old cast-off pieces of the past. He thus keeps the traditional process alive as well as saving the products of previous artisans. On a number of occasions he has said, "If I find nice piece of old iron, I'll save it. I like to study these old iron. It's important to keep it."

Plate 18

46 MURRAY BOULEVARD

In 1952 Philip had to move his shop to Laurens Street because the property that he rented on Alexander Street was sold and cleared to make way for an electronics company. The new location was again near the waterfront, but his clientele remained mainly homeowners in need of architectural embellishments. Proof of this continuing demand is seen in a large driveway gate at 46 Murray Boulevard.

It is a very large gate, sixteen feet wide and more than nine feet tall, with two symmetrical swinging sections. The top of the gate has an elaborate assemblage of **S** and **J** curves of various sizes, all set in a triangular configuration peaking at the point where the two gate sections touch. The screen section is divided by a horizontal bar of flat iron with the upper unit twice the size of the lower one. A twisted bar runs through the center of each half of the gate and is surrounded by large **J** curves paired to form a heart. Each scroll is coiled into two and a half turns. To each side of this heart-shaped motif are three smaller **J**

18   Gate, 46 Murray Boulevard.

scrolls and two **S** scrolls. Two small **J** curves were placed over the upper heart, while two **G** curves support the lower one. The lower third of the screen has many of the same motifs used in the upper section, with many small **J** curves added to fill the space. This dense arrangement of ornament serves the function of dog bars (that is, keeping out small animals), but with a sense of grace. Since dog bars are usually straight bars, the use of scrollwork may be considered an indication of growing importance of the ornate in the career of a blacksmith trained as a wheelwright. While he

deferred to the customer's request to use curved pieces when he wanted to use bars at 2 Stolls Alley, in this case his design was his own. He says:

I've never made anything I didn't like. I take a long time on the drawing, and when the customer likes it I already liked it first. I'm lucky his choice is mine too.

Seventeen years into his decorative venture as a blacksmith, Philip was well on his way to becoming an artist.

Plate 19

125 TRADD STREET

This driveway gate is very tall and relatively plain. Its two halves consist of straight bars with a surrounding border of **S** scrolls. These are aligned singly along the sides and are doubled at the top and bottom to form lyre shapes. Above the screen two **J** curves and an **S** scroll are arranged in an ascending line from the outer edge to the middle bar of the gate. Interspersed in this assemblage are four **J** scrolls paired in such a way that they seem to be two small **S** curves which penetrate the larger scrolls.

Plates 20–23

100 TRADD STREET

Down the street a few blocks from the previous work are two walkway gates, a single and a double gate. The designs of both are the same, but they are made in dimensions appropriate to their different widths. Over the top there is a row of square bars beaten into lance points. Every other spear has two **J** scrolls welded to it to form a fleur-de-lis. The gate screen is mainly straight square bars except for a lunette section at the top and the lock bar ornament at the middle. In these two units, several **C**, **S**, and **J** curves are combined to form a unique design.

Consider the larger gate (plates 20, 21) first. There we find two **J** curves paired and the end

19   Gate, 125 Tradd Street.

20   Gate, 100 Tradd Street.

22   Side gate, 100B Tradd Street.

21   Detail of gate, 100 Tradd Street.

of the longer scroll hammered into a wavy
taper so that it undulates toward the end of
the lunette. Next, two **S** curves are notched
together with leaves and **J** scrolls welded to
their point of intersection. Finally, two **C**
scrolls are notched together. The lock-bar
decoration has the same **C** curve combination
as a center element which is flanked by
asymmetrical diamond and scrollwork pieces.
An identical set of scrolls is used to decorate
the lock bar in the smaller gate. However, the
lunette section of the smaller gate (plate 22)
varies significantly. We find it filled with

slams against the strike bar and falls back into place by gravity (plate 23). Since Philip usually installs this kind of lock, the box lock on this gate is, for him, a further extension of his metalworking ability and therefore he regards this gate as special.

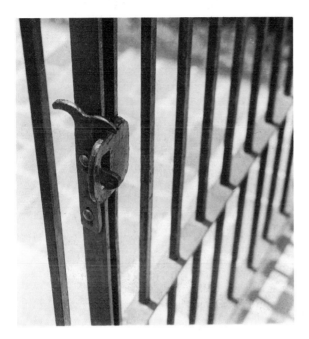

23   Detail of a "Charleston latch".

interlocking **C** scrolls flanked by interlocking **S** curves with a waved taper at the corners of the gate. These waving elements also project from the centers of the pairs of **S** scrolls.

Beyond the unique set of decorative elements, this gate is important because Philip also fabricated the lock mechanism, which is housed in a sheet-metal box resting on the lock bar. With justifiable pride Philip notes, "It still works, too." Although key locks and brass knobs are seen on many Charleston gates, it is more common to use a simple slide latch that rises automatically when the gate

Plates 24–27
43 WENTWORTH STREET

St. Andrew's Lutheran Church was rebuilt in 1838 with three large gates under its impressive Greek Revival portico and a wrought-iron fence with a smaller fancy walkway gate set to each side of the building (plate 24). In the early 1960s the church board acquired adjacent property that required a wrought-iron gate, and they wanted to keep all of their ironwork the same. The board offered the job to Philip. Here his imagination was severely shackled, since his task was to replicate rather than create. Yet there was challenge in this job since he had a fine work of the early nineteenth century as his model and could now prove that he was equal to the old iron masters. Since St. Andrew's is a historic building, his clients were very cautious about Philip's work and took extensive measures to make sure that the new gate (plate 25) was an exact duplicate of the older walkway gates. They so wanted the twentieth-century work to look as if it had been there for over a hundred and fifty years that they even demanded certification that wrought iron and not mild steel was used. This detail Philip easily accomplished. More

25   Gate, 43 Wentworth Street.

24   Nineteenth-century gate, 43 Wentworth
Street.

26   Detail of gate, 43 Wentworth Street.

27   Detail of gate, 43 Wentworth Street.

important, he proved his skill as an ornamental ironworker.

The gate screen itself is very plain with a band of scroll-and-leaf decoration across the lock bar (plate 26). The gate swings between two elaborate pilasters filled with pairs of **S** and **C** scrolls with the larger ends of the **S**'s shaped into elliptical volutes (plate 27). The pilasters are crowned by a lyre form embracing a flat bar that ends in a wavy taper. Over the gate is a high overthrow composed of very large **S** and **J** scrolls. Between them is an arrangement of flat bars which outline a diamond and terminate in an iron sphere and a pointed wavy finial. Scroll for scroll and bar for bar, Philip's gate is the perfect match for its older mates. He thus satisfied the church board, in addition to answering any questions about whether his work was equal to the standards of the Charleston tradition.

Plates 28–31

61 CHURCH STREET

Another request for antique-looking ironwork was made by the directors of the First Baptist Church, a structure built in 1822 after the design of architect Robert Mills. He is thought to have designed the wrought-iron gates and fencing (plate 28), as well as the heavy Doric-style building. Philip was asked to design and install a fence at the back of the church alongside the graveyard, and a double gate closing an arched passage through the church school to a playground. While the new

*28   Detail of nineteenth-century fence,
61 Church Street.*

has a pattern featuring loops that interlock
with a central circle, outline a diamond, and
end in matched curls (plate 29). This is the
same pattern Mills used to ornament the
pilasters of the front fence, except that Philip
has used it horizontally instead of vertically.
He says of his work:

I wanted to stay within what they had on the
front. I had like this [design]. So instead of a
pilaster I put it in a section and balance it up.
In fact I told them I would try to stay within
what they had as much as I can.

   In the pilasters of the graveyard fence (plate
30) the diamond element was eliminated.
Philip used instead a simpler arrangement of
interlocking loops with curled ends. The
diamond motif does occur on the lock bar of
the graveyard gate, but only once, while in
Mills's design it occurs three times. Philip's
fence fits harmoniously with the old design,
since he used the same dimension of round
pointed bars and the same Y-shaped braces
with scrollwork ends. The gas lamps that sit
atop his fence (plate 31) are held in place by
lyre-shaped standards comparable to, but not
as fancy as, those used in Mills's section of the
fence.

ironwork had to match the old, it did not have
to be an exact duplicate. When Philip made
the required gates and fences, he managed to
express his own sense of design while
following the lines of the established
nineteenth-century motifs. This was very
evident in the gate in the school, which is
largely a plain arc-shaped grill of square bars.
But in the corners of the opening there are
brackets with **S** and **J** curves, and the lock bar

Plates 32, 33

34 MEETING STREET

The gates that Philip made for this residence
derive their design in large measure from the
decoration done for the First Baptist Church.
Both the drive and walkway have plain bars
with an ornamental lock bar featuring a
central diamond, and flanking curves that end

29   *Detail of school gate, 61 Church Street.*

30   *Graveyard gate, 61 Church Street.*   31   *Lamp standard, 61 Church Street.*

32   Detail of driveway gate, 34 Meeting Street.

33   Garden gate, 34 Meeting Street.

with a semicircle. This same design is
repeated twice in the overthrow of the
driveway gate. Both gates are tipped with
barbed points made by welding short pieces of
flat iron stock to each of the square bars. The
top of the overthrow is decorated by an
assemblage of **S** and **J** curves, and is crowned
by a spire that Philip calls a "wiggletail."

   The house at 34 Meeting is thought to have
been built in 1760. Although it had handsome
wrought-iron stair railings, its gates (at least
during the first half of the twentieth century)
were a humble wooden variety. By means of
Philip's skill, this pre–Revolutionary War
house has been graced with decoration
befitting its age and stature.

Plate 34

1 LEGARE STREET

Set between two ivy-covered brick pillars is a
garden gate with one of Philip's favorite
designs, the "circle scroll": "It's tied to historic

34   Gate, 1 Legare.

Charleston architecture, and that's one reason
I like to do it. It keeps part of an old tradition
alive and it tells us something about the past."
The circle is divided into six sections with two
J curves in each one. Each curve radiates like
a wheel spoke from the center rosette and
ends in a curl at the edge of the circle. The
scrolls are curved so that they appear at a
glance to interlock and form a six-petaled
flower. The corners of the panel with circle
scrolls are filled with **C** curves, and the whole
gate is topped by a pair of **S** curves with an
outlined spear point between them. The lower
panel of the gate is a grill of flat slats set into
a diamond lattice pattern which, in Philip's
opinion, looks as if it "belongs on a jail."

Plates 35, 36

329 EAST BAY STREET

The gate at the Thomas Gadsden house is
important for several reasons. First, it is a
large driveway gate with an impressive
rattlesnake emblem. Second, the snake motif
commemorates Gadsden's invention of the
"Don't Tread on Me" flag used in the
Revolutionary War, thus highlighting a bit of
local history. But more important is the fact
that this gate is the first work in the long
tradition of ornamental wrought iron in
Charleston to feature a sculpture of an
animal. Philip made here a distinct and
unique contribution with this piece, which

35   Gate, 329 East Bay Street.

36   Detail of gate, 329 East Bay Street.

initiated a new chapter in the saga of Charleston blacksmithing.

According to Philip, the Historic Charleston Foundation, in the person of Samuel Gaillard Stoney, a prominent historian of Caroliniana, commissioned the gate. "He knows the story, that's how come the snake. . . . No, they didn't give me a plan. They know what they want. I just drew it out, a sketch." The fabrication of the whole gate went rather quickly. In a few days the frame and bars were welded in position. The lock bar was decorated with three diamond and scroll units, and the top lunette arc was filled with interlocking **S** curves with waved centers, plus other **C, J,** and **S** scrolls. A series of lance points was welded to the top of the frame. Each point was forged into an elongated diamond shape with a ridge running from its base to the tip. The snake was the last element made for the gate. It is a boldly curving serpent turned into four loops. Its head is broad and arrowlike with a protruding forked tongue. The tail shows seven rattles, which Philip shaped with swaging tools. He declares, "That's the age. Every year he get a rattle, don't he."

The forging of the snakes for the two halves of the gate consumed as many days as were needed to build the rest of the gate. Philip remembers most of the details.

We made the gate in about eight days, you see. I think we been almost as long on the snake as we been on the whole gate. I start with a piece five-eighths inch by two and one-half inches. That's big metal. I think I went to the junk yard and pulled it in. You heat and beat, heat and beat, heat and beat. That's the name of the game. The more you look down, the more you see. It's like you never goin' finish. I been just as long on the head as I been on the body of the snake.

There was a complicated thing about it, putting the eyes in him [so] he'll be looking at you when you go there. We had to put the eyes in it. We judge it. On the first sight I made the snake, the body. But the eye, I had to make several changes. Like the first, second, third, the fourth time I placed the eye in the head of the snake, it look at me. That was the snake. The eye was complicated. You put the eye in it and you just see something that look like a dead snake. He look dead, you know that's true. You got to get that eye set that he look live. You look at it now sitting up there in the gate. Now you see it looking after you anywhere you are, any side you on, looking dead at you. That's the snake. That's the important thing about the snake . . . to look live. That's all. He [Samuel Stoney] didn't want a dead snake. It's a real rattlesnake. It's in a coil. He was ready to strike too.

The tongue weren't too hard. Take a little thin thing. Very delicate, you see how it look. I made that and put it in his mouth. That's separate. Move it around until he fits just like the eye.

It is evident from Philip's story that this gate is very important to him. He regards it as one of his best, maybe even at the top of his personal list. It was an adventure to create a snake and he recalls, "It's more than just beatin' it out. I enjoyed it and I got of experience from my work." Since this gate is usually kept in its open position, Philip's work is seen only by those who seek it out. Perhaps as the concern for local history expands, more Charlestonians will want to see the Snake Gate at the Gadsden house.

At the same address is a second gate that Philip made. Seen next to the Snake Gate, it would surely be overlooked. It was not meant to be fancy, even though it is part of an impressive three-story brick planter's house. It does have, however, some of the same decorative motifs used in the Snake Gate. It has the same square bars tipped with beveled lance points and a lock bar ornamented with scroll and diamond forms. It is basically a narrow but tall walkway gate giving admittance to a side passage to the house.

Plate 37

56 SPRING STREET

Spring Street is in the uptown part of Charleston, far removed from the famous promenades and alleys of the Battery. It runs through a working-class neighborhood of plain, frame houses owned or rented almost exclusively by blacks. Philip has done little fancy work in this area, since only the comparatively wealthy are motivated to install ornamental wrought iron on their houses and, more important, have the money to pay for it. In Philip's broad categorization of clients, there are "rich folks" and "poor folks," and the people on Spring Street fall usually in the latter category. The gates that he has done in this area since the mid-1960s are thus special instances in which a certain individual's desires have overwhelmed Philip's business sense. This ornamental ironwork has been installed for less than normal prices, and yet it is as decorative as any found in the city. It is almost as if the emotional appreciation Philip received caused him to outdo himself.

So for two hundred dollars, an amount his client saved for years to accumulate, Philip did his best to satisfy not only her, but himself: "I think this is a nice piece of work. That's what she wanted, and that's what she got."

The ironwork consists of a low driveway gate and short stair railings alongside the porch entrance. The basic decorative unit is a lyre form, which occurs in the screen of both halves of the gate and in between the balusters of the railings. The decoration in the gate consists of large and small pairs of **S** scrolls and **G** curves. A twisted flat bar runs through the center of the motif. The dog-bar part of the gate is divided into sections, the upper one filled with overlapping **J** scrolls plus smaller **S** curves and the lower one with elongated **S**'s that seem to interlock at the center. Over the top of the gate is an arrangement of scrolls Philip commonly uses in simple gates: a large **S** runs from the center of the gate to the edge with two small **J** curves attached to the middle of the large scroll in a manner resembling an interlocking **S**. In addition, the flat bars at the middle and edges of the gate are curled into matching curves. The hand railings also terminate in scrolls and are topped with brass knobs, which Philip made from scratch. When shown a photograph of this gate, Philip smiled.

I glad you got something uptown. People think that all my work is in the lower Battery. But I done a lot of work for poor peoples like myself. They enjoys it too.

37   *Gate, 56 Spring Street.*

*38   Gate, 254
Coming Street.*

Plates 38, 39

254 COMING STREET

Around the corner from the Spring Street gate and up two blocks on Coming Street, Philip did another commission that is somewhat out of the ordinary. There is a short fence of plain pointed bars, a walkway gate, and a driveway gate. The gate screens are for the most part plain, just like the fence sections. Their lower panels feature combinations of **J** curves set to look like interlaced **S**'s. The top of the driveway gate is indistinguishable from the gate at 56 Spring Street, and the top of the walkway gate uses the same design found at 1 Legare Street. This ironwork is not then remarkable except for the use of a sculptural emblem to indicate the identity of the customer. Set in the center of each gate is an iron upholstery hammer. It is an accurate, life-sized imitation of the tool used in the furniture-covering business to set and hammer tacks. Philip did not acquire the hammer heads and then add iron handles, but rather he forged the entire object out of the same bar stock used in the rest of the gate. When they were finished he cut out sections of three bars and welded the hammers in the space.

The hammer is meant as a humorous statement, a pun on the identity of the customer. Usually, initials are used in a center panel, but since the owner of the house was an upholsterer, Philip thought it a unique chance

39   Detail of gate, 254 Coming Street.

40 *Fence and gate, 55A King Street.*

to show off his old toolmaking skill. At the same time he enjoyed the humor of his design. When viewing this gate together, we encountered three elderly men passing the time of day in the next yard. Philip knew them all by profession, and in order not to tarry too long in idle conversation, he directed their attention to his gate: "When I make a gate I put an initial or mark it. Upholstery man, I put a hammer. Blacksmith, I put a sledge or anvil. But you, the plumber, I put a toilet bowl."

Plates 40, 41

55A KING STREET

Set well back from the street and hidden by thick shrubs is the most ornate of Philip's fences. This low garden fence has tightly spaced rods, each topped with a lance point and a pair of half-scrolls. The diminutive size of each spear makes it seem more like a floral motif than a defensive device. The gate top has two large J's, each adorned with two leaves, and a large spear point decorated with small scrolls. The lock-bar section has the same pattern of S's and leaves as the top of the gate, only both ends of the scrolls are of the same dimensions. The lower decorative panel features J curves arranged to look like overlapping S scrolls. This pattern is similar to that used at 68 Alexander Street and 254 Coming Street, but features four J scrolls instead of two.

41 *Detail of fence, 55A King Street.*

42   Gate, 28 Queen Street.

Plate 42

28 QUEEN STREET

This is a deceptively simple gate because the screen is mainly plain bars. But the top ornament and dog-bar section show an inventive use of standard elements. The top portion of each gate section is chiefly a large **S** scroll which creates a space between it and the top of the screen. This arc is filled by a large **J** curve and is matched by a small **J** scroll on top of the **S**. A similar **J** curve sits between the large **S** and the center post. The base of the gate features a combination of **S** curves. The largest one fills in the two sides with a long straight diagonal between. Small **S**'s are set on each side of this slat so that their larger volutes touch the ends of the large **S** scroll. The small ends of these smaller **S** scrolls overlap so that they are set just opposite the larger curls. Even though there is no central axis, the whole design is rigorously symmetrical, because Philip has divided the diagonal into thirds and placed the overlapping **S**'s in the middle third. Of course, when the gate is closed the design is clearly symmetrical, since the halves are mirror images of each other.

Plate 43

72 ANSON STREET

In 1969 Philip moved his shop for what he hoped would be the last time. He chose as his new location the yard behind his house at 30½

Blake Street. The Laurens Street shop had been taken over by the Port Authority to make way for expanded facilities. He felt that if he set up in his own yard he could never again be forced to move. His clientele followed him to his new shop, and so his business did not falter at all. Many of Philip's customers were now building contractors who specialized in the restoration of historic houses. Many properties in the Ansonborough section of Charleston have been restored, and consequently their owners have had need of Philip's services. Indeed, a three-block stretch of Anson Street features his work from end to end. At number 72 a long fence separates the house and yard from the street. At first the fence seems standard enough, but closer inspection reveals that above the top horizontal the bars taper gradually to a point. Often the point on fence bars consists of an abruptly angled tip covering an inch or less. In this case, the last six inches have been drawn down into graceful spearlike tips over two hundred times across the width of the fence. Countless hammer blows were required with repetitive heating of the iron. Hours of sweaty forge work were needed to produce what seems to most no different from factory production.

To Philip the fence is just right for the house because "it looks like an old piece. Eighteenth century. [*Why?*] The size of the iron and the way it been worked and hammered out." His task with this commission was to provide a historic fence for a historic house. The only emotion that it evokes for him is the memory of tremendous effort. The blacksmith's effort is not appreciated unless the results are elaborate. If the work is plain, even though painstaking, few will care and fewer still are inclined to pay for it at a rate

43   Fence and gate, 72 Anson Street.

equal to the difficulty. In Philip's view, "those bars put a whuppin' on you."

The gate in front of the house has the same simple bars as the fence, but with a lock bar decorated with two **S**'s and a central ring. The ring motif is also used in the two flanking pilasters, where it is combined with lyre shapes with wiggletail centers.

Plates 44, 45

74 ANSON STREET

The next house on Anson Street also features Philip's work: a low fence with a fancy walkway gate and stair railings leading to the piazza. The gate itself is not especially

44   Gate, 74 Anson Street.

45   Newel post,
74 Anson Street.

unusual, but its two pilasters are. They are filled with two **S** curves, which are as wide as the pilaster, set above and below a pair of **C** scrolls. These panels are topped by small lyre forms and by the same outlined spear point found on the gate. The railings consist of a plain flat bar with square balusters. They end in a full horizontal spiral, a difficult maneuver when forging a flat bar. The newel post is topped by a large iron knob (plate 45).

Plates 46–50

45 MEETING STREET

The large house known as "Eagle Nest" at 45 Meeting Street is rich with Philip's decorative work. There is a curving stair rail with elaborate newel post, the porch landing rail, and three different sets of window grills. All of this work was done with the aid of modern tools—in particular, an electric welding machine. Yet the contemporary technology does not interfere with the nature of the nineteenth-century designs that he used. Philip employs modern means to preserve an old art form.

The railing follows the line of the steps, which start at the right edge of the porch landing and then curve slightly toward the left as they descend to the walkway. The railing lacks any decoration, but the newel post is spectacular (plate 46). The thick bar stock coils through two full turns and is topped by a faceted polygon finial. No less eyecatching is the central panel of the porch rail, which displays a fluted vase surrounded by a fringe of **S** scrolls (plate 47). The vase form

46   Detail of newel post, 45 Meeting Street.

is composed of bent slats, tiny rings, and **J** curves. Vines formed by a chain of **J** scrolls run from the top of the vase down its sides to its scrollwork pedestal. As Philip pointed out, "Everything balance, it's not top heavy." In design, this motif is very similar to an urn used in a gate at St. Michael's Episcopal Church done in the mid–nineteenth century by J. A. W. Iusti (plate 48) and to another urn decoration found in the main gate to Boone Hall Plantation just outside Mt. Pleasant. Philip acknowledged that his client gave him a plan to follow, but added quickly, "They gave me this design, but we made a lot of little changes." Thus the final pattern is simultaneously emblematic of Simmons and Charleston. The porch railing design is also used for the window grills on the first floor of the house (plate 49). In the second-story windows (plate 50), square ironwork panels are divided into three sections: the top section consists of a band of rings linked by short horizontal bars, the large central

47   Porch landing, 45 Meeting Street.

section has five vertical loops interspersed with lead rosettes, and the bottom band is filled with rectangular fretwork. The two rectangular grills on the dormer windows (plate 50) are also divided into three sections. The uppermost unit is a band of symmetrically arranged **S** curves and contrasts with the bottom section, which is a band of interlaced diagonals that outline diamonds and triangles. The center section features interlocking circles.

There is no doubt that all of these designs are complex and intricate, but the production processes were also complicated, requiring close control over the multitude of small parts and careful attention to their assembly. One of the dormer grills alone required almost one hundred separate welds. The craftsman's need for precision does not decline with the use of modern labor-saving technology. A blacksmith is still accountable for his work, and the pieces at 45 Meeting Street show how much credit is still owed to the artisan even when he employs modern machinery.

48   Detail of St.
Michael's graveyard gate.

49   Window grill, 45 Meeting Street.

50   Upper window grills,
     45 Meeting Street.

Plate 51

225 EAST BAY STREET

The old open-air market once bustled with street vendors who bought vegetables and hustled them throughout the city. Philip's first work as an apprentice often consisted in the repair of their push carts. In the 1970s the market area was once again a place where his effort and skill were required, although now in a decorative and more permanent manner. One of the waterfront unions has an office building with a parking lot behind it which is entered from Market Street. Just twenty feet from the colonnaded shed where Philip's wagons once rolled, a broad driveway gate that he designed now secures the property.

It is mainly composed of square bars with an inverted arc at the top. This arc is decorated with **S** and **J** scrolls. Part of this ornamentation is a general pattern used on driveway gates of conventional width. The extra length in this gate was thus filled by a combination of the usual decoration plus some added fillers. Other special touches include the series of paired **C** scrolls throughout the lock bar and the rings at the base of the gate.

51   Gate, 225 East Bay Street.

The locking mechanism on this gate is unique in Philip's repertoire, for in addition to the usual Charleston latch-and-rod stop, here he used a lever-action catch that automatically holds the gate in the open position when the gate swings against it. The catch releases when it is depressed by one's foot, and the gate is simultaneously pulled forward.

Plates 52, 53

78 EAST BAY STREET

The four window grills and doorway-arch panel on this house were created by Philip when his customer could not quite say what she wanted.

52   Doorway grill, 78 East Bay Street.

53   Window grill, 78 East Bay Street.

The whole house, all those window grills, that's my design. I draw that. Sometime they request. They could try to explain to you what they want. And you sit down. You say, "They want it to [look] open, but I'll make . . . I'll draw this one. See how they like it."

The window grills are squares sectioned into quadrants, each filled with paired **J** curves embracing a wavy center. The four sections fit together in such a way that a fleur-de-lis is formed at the center of each side with its center pointing toward the middle of the grill. The quadrants also form a four-petaled flower. The arch section over the front door is filled with **S**, **G**, and **J** curves arranged symmetrically on either side of an outlined diamond.

Both of these compositions are important to Philip as designs rather than examples of skill. He regards his creative ideas as unique achievements. This was revealed a few years later when a woman living on Tradd Street, just around the corner from 78 East Bay, wanted some window grills made for her house.

The lady, I took her around there and show her this. When I show it to her, she say, "That's exactly what I want." I said, "But I can't make it like that unless you ask the people in the house 'cause it too close together, right around the corner." But I told, I said, "I tell you what I'll do. I'll put a rosette there. That's makes it different." That's the joy I get out of doing ironwork.

Most passersby would think the window grills at 18 East Bay and 10 Tradd were the same design, and they would be correct. But in the eyes of their creator, a minute difference is enough to make them distinct works, each one with a unique biography. For Philip, like many contemporary artists, the design process is sometimes more significant than the final product.

Plates 54—56
4 LEGARE STREET

When the house at the corner of Legare and Lambol Streets was renovated in 1972, Philip was given a contract to provide a variety of ironwork ornaments. The most prominent was the balcony which hangs over the front door (plate 54). In overall form it is rectangular, with a curved central panel extending an extra two feet away from the house. It thus matches two original balconies on the south side of the house. Supported by two **S** brackets, its balusters are straight spokes which are alternately curled at the ends. Each spoke is decorated with a lead rosette. Over the driveway is an arrangement of **S** curves which embrace the outline of a spear point (plate 55). Plain railings guard the front steps. The most attractive decorations are small grills set into the wooden doors that lead to the front walk (plate 56). The scrolls in these insets are made from small square bars and consist essentially of a pair of **G** curves with **J** scrolls attached. The whole unit has the look of a swirling diamond form with a long split tail attached to it. The same square bar stock is used above and below this motif to fill up the space. These bars were twisted so that they seem more like strands of rope than pieces of wrought iron.

Plates 57, 58
107 TRADD STREET

Not all of the decorative blacksmithing of Charleston is visible from the street; much of it is indoors. A splendid example of the kind of design Philip can produce in the interior of a house is the stair and landing railing at 107 Tradd Street. The staircase bends around the back and side walls of the foyer. Flat bars serve as the base and hand rails, and square

54   *Balcony, 4 Legare Street.*

55   *Detail of driveway gate, 4 Legare Street.*

56   *Detail of walkway gate grill, 4 Legare Street.*

57   *Detail of stair railing, 107 Tradd Street.*

58   *Detail of landing railing, 107 Tradd Street.*

bars as balusters. This same square stock is also used for the two decorative panels, one in the long section of the stair rail and one in the long section of the landing rail. The dimensions of the ornamentation thus match the structural components, subtly blending both together.

The first panel has an asymmetrical leaf pattern resting on a **C** scroll between two large **S**'s. The **S** curves both face the same direction but have their ends reversed so that the large curl is opposite the small one of the other piece. The landing panel uses the same elements, but in a slightly different order. Here there are three leaves, the central one touching the hand rail. The large **S**'s outline a lyre pattern and thus provide a balanced frame for the asymmetrical leaf cluster.

Philip refers to his leaf outlines, which roughly approximate curved parallelograms, as "magnolia leaves." He explains further:

I had to call them something, give them a name. They seemed a little bit like magnolia leaves, so that's is what I called them.

While unique in shape, they also resemble the spear-point outlines with which he often tops his gates.

59   *Gate, 2 St. Michael's Alley.*

60   *Detail of gate, 2 St. Michael's Alley.*

Plates 59, 60

2 ST. MICHAEL'S ALLEY

This walkway gate is another example of Philip's sculptural ability. Set in an oval frame in the upper portion of the gate is the outline of an egret. Seen in profile, the bird's

head and beak point skyward while it seems
to raise one long leg in order to take a step.
Philip's rendering is very meticulous. He
includes details of the knees, talons, tail
feathers, and beak. Although the explanation
he gives for this sculpture awards a large
share of the credit to his client, we can still
understand how much of his own design skills
were involved.

That's what Mr. Rhett had wanted. He asked
me could I make it. I said, "Well, bring me
some kind of drawing [and] I'll make it for
you." He went back and he got me a plan. It
just give me an idea of what they want. What
he wanted. Then I make an accurate drawing
and then I cut a pasteboard out. He give me
a bird and I have to make it full size. He show
me, he got a picture and from there I arrived
that [bird], size and everything.

In the upper panel of the gate the bird rests in
its oval frame as a vertical composition which
is further framed by a horizontal oval. The
spaces between the two ovals are filled with **J**
curves and the corners of the panel are
decorated with **S** scrolls. The lock bar also has
S scrolls and the initial **R** standing for Rhett,
the family name of the household. Above the
gate screen is an arched overthrow bearing a
palmette design shaped by **J** scrolls and loops.
The palmette is flanked by pairs of **S** scrolls
connected by an arc of flat stock.

*61 Gate, 67 Broad
Street.*

C's and end in wavy tapers just short of the center scrolls. This same complicated pattern is repeated in the lock bar of the gate, which has barbed dog bars and small **J** scrolls at the top of the gate screen. The most intriguing part of the gate is the overthrow. Here, pairs of **G** scrolls have been interlocked in reverse directions. A pair of **C** scrolls fill in the space between the **G** curves, while **J** scrolls decorate the ends and top of the arch. In the middle of the panel, near the top, three wiggletails come together, forming an interesting negative space composition.

62   Detail of gate, 67 Broad Street.

63   Ventilator grill, 67 Broad Street.

Plates 61–63
67 BROAD STREET

The ventilator grills and gateway of this commercial building, made in 1974, employ designs with eighteenth-century analogs. The grills are some of the most elaborate for their purpose to be found in Charleston. At their centers are a pair of "broken scrolls" from which **J** curves extend to each side. Two bars reach from the edges of the grill toward the

Plate 64
THE STAR AND FISH GATE

In 1976 Philip, along with his coworkers Si Sessions and Ronnie Pringle, were invited by the Smithsonian Institution to demonstrate their blacksmithing talents at the Festival of American Folklife in Washington, D.C. For two weeks in August, on the mall near the Lincoln Memorial, Philip designed and constructed the Star and Fish Gate. It was intended to display a wide range of forms in order to explain to festival visitors just what he did in Charleston. Consequently, the motifs are combined in unusual ways. At the top, **J** curves were inserted inside **S**'s. Two of the spokes were split and opened. The central panel, bracketed by crescents and flat **C** scrolls, contains a spectacular star within a star. The lock bar has two **S** curves adorned with pecan leaves. In the lower section there

64   Star and Fish
Gate.

is a fish, more particularly a spot-tail bass, outlined with slats (the same technique used to fashion the bird at St. Michael's Alley). Philip notes:

Fish could have been cut out of one piece of metal, but using several pieces, you create an effect. It's open. You see there [in the lock bar] you got this [open] effect and you get it here with the fish too.

The open effect is a subjective measure of the gate's beauty. Yet it is clear that however beauty and attractiveness are assessed, this gate is a handsome design. In fact, it has been adopted as a cover design for the *Directory of South Carolina Craftspeople*, a listing published by the South Carolina Arts Commission.

Plate 65

16 BEDONS ALLEY

The metalwork insert for this wooden gate was executed by Philip and follows a very precise blueprint which was provided by his client. The central motif is a shield bearing a chevron topped by a crown. This unit is set inside a circle which is encircled by a second ring plus a fringe of **J** scrolls. The inner ring is attached to the outer one by three **X**'s. Philip's only deviation from the original plan was to shift the two lower **X**'s toward the bottom just a bit. In this way the supports took on a vertical line. Philip thought that this would look better and his client apparently agreed.

65   *Gate, 16 Bedons Alley.*

The Significance of Changing
Market and Technology

The foregoing portfolio of works gives us an indication of the direction of Philip's career as a blacksmith over the last forty years. In these pieces we can trace the growth of his reputation and the development of his skills. He was, to be sure, already a competent craftsman when he made his first gate at Stolls Alley, but by the time he completed the Snake and Bird gates his sense of artistry had truly expanded. Over the years his repertoire of decorative items increased in range and content. At first he made different kinds of gates, but then he added balconies, window grills, stair railings, and even interior decoration—in short, all manner of architectural embellishment.

The flourishing of Philip's career as an ornamental ironworker is tied directly to the activities of the Historic Charleston Foundation. Beginning in the late 1930s, the Foundation has promoted with increasing enthusiasm the restoration of old properties and hence has helped preserve many old professions. Ornamental ironworking is among these trades, along with masonry, cabinetmaking, plastering, and others. Philip is very specific about what he calls the "Historical Foundation Society."

First everyone destroy the old architectural structure of the house. When they keep the value and they keep it then finally they demand it. Then it become a compulsory that you don't tear down an old house or tear down an old wrought-iron fence. It become a compulsory and that's what happen. You see, I could . . . we had the forge, start on the forge and you can roll them things just like the old craftmen used to do. The young people [ironworkers] then they didn't do it. They went the fast way. So we stop, stopped right there and pick up all the work. That thing really helped.

By retreating to the past Philip assured his future in Charleston. As the local citizenry's fondness for old buildings grew, so did their dependence on craftsmen with knowledge of old ways. Philip's ability to make old-fashioned artifacts ironically earned him an active contemporary market.

While his gates and grills continued to look old, Philip updated his tools and techniques. One of his first changes was to put an electric-powered blower on his forge.

When I went from Alexander [Street] to Laurens [Street] that's when I become more modern. I convert the hand forge on Laurens Street. When I was an apprentice we had a hand blower. The Old Man, Peter Simmons, never seen an electric forge because he went with the hand forge. After I come and took over the supervising part, the helps then they were picking their work, and if want help then you got to make it much easier. So I convert the hand blower into electric blower.

In the early 1950s he acquired an electric arc-welder. This machine was used primarily in car body work and small "fix-it" jobs, but it was also used to speed up the assembly of wrought-iron gates. Philip's earliest gates were put together with rivets.

We got a rivet tool. One hold the tool against the head of the rivet and I take the riveting hammer and rivet it, flatten it out. Take us three times longer than welding.

Although there clearly was an advantage to the new machine, in Philip's view it was hardly different from the older process.

It's no different because if you do the riveting you've got to grind it off and make it look neat and the welder you grind it off. Same principle. It's just one's faster than the other. The rivet is a slow way of doing it. The rivet is a slow way and the welding is a fast way, let's say that. 'Cause you do it with the rivets you got to make it look like the weld. You couldn't tell the difference. Some of the rivets, the head is riveted in so tight you couldn't see the head, see the rivet at all.

Almost all the work that came out of the shop on Laurens and then Blake Streets was arc welded. Philip's most substantial works, those which display the highest degree of competence and creativity, were all welded together. The elaborate grills at 45 Meeting Street, which so strongly echo the nineteenth-century designs of St. Michael's, were fashioned with the same technology as modern skyscrapers and ocean liners. Philip can create old forms with new techniques or use the modern means to develop new expressions. His bird and fish sculptures were easy to fabricate with his arc welder.

Occasionally a client will request extremely antique-looking ironwork which requires Philip to use his older blacksmithing methods.

And some work carries a rivet with a head. Some people want to see it. You do work now sometimes, I got to use the rivet because especially if you working on a piece of wrought iron, an old piece that have the rivet in it, then I got to do it with rivet. That's the slow way of doing it, but you have to do it.

The skills that were once essential, such as riveting or forge welding, can now be used solely for aesthetic reasons; instrumental technique is converted now into expressive detail. Philip's modernization of blacksmithing is thus a complex process, since the changes and improvements achieved by using new tools and equipment are selectively chosen. The changes were not the inevitable result of the passage of time. Traditional forms can be preserved or ignored; the traditional techniques can be used or set aside. Philip is still very much in control, regardless of the mode of his performance. The blacksmith, the worker of the metal, presides over the changes in technology. The more Philip adopted modern tools, the more he was able to meet the local demands for decorative ironwork and thus fulfill the purpose for which he was trained. It is commonly assumed that the use of contemporary technology sounds the death knell for traditional crafts, but we see in Philip's case that his professional life has flourished with new modes of metalwork. The availability of modern tools, techniques, and materials has made his work methods more complex, but his works, in general, still flow in the mainstream of the Charleston tradition.

The Charleston Tradition

As early as 1739 Charleston houses featured wrought-iron balconies. These seem to have been architectural afterthoughts added to help make British houses more pleasant in the subtropical Carolina climate. Balconies were the major form of decorative ironwork made in the city before the Revolutionary War. However, the first colonial blacksmith to advertise his ornamental abilities, James

Lingard, claimed in 1753 to "make all kinds of scroll work for grates and stair cases." We can conclude then that local smiths were trying their hands at a variety of forms in order to satisfy the demands of wealthy planters and merchants for elaborate mansions and gardens.

Their designs were mainly imitations of British works or derivations from published plan books. The altar gates and railing at St. Michael's were imported from England in 1772, and provided inspiration for at least a dozen window grills, a balcony, and the portal gate at St. Philip's. A 1765 pamphlet entitled *The Smith's Right Hand; or, a Complete Guide to the Various Branches of All Sorts of Ironwork Divided into Three Parts* is mainly a collection of plates depicting many gates, fences, railings, and balconies. Pattern books like this one found their way into Charleston smithies and no doubt had a considerable influence. Indeed, the book just mentioned has several plans for fences and gates found in Charleston. While South Carolina remained a British colony, its art forms were decidedly English, and London styles set the tone for Charleston. Even into the late eighteenth century, British patterns shaped the local preferences in wrought-iron decoration.

When new designs were attempted, they were often mixed with older patterns. The railing on the William Gibbes House has a distinct center panel, but the support panels under the lamp standards are derived from St. Michael's altar rail. Early Charleston ironwork has been criticized as lacking the ornamental virtuosity of its British antecedents. The smiths have usually been excused, however, because of their lack of training or poorly equipped shops. But the mixture of forms as found at the Gibbes House suggests that a new regional style was being developed, that Charleston's ironworkers were searching for a unique set of artistic expressions. Their new repertoire of forms would be based on historical precedents, while new motifs were combined with them in unexpected ways. Charleston blacksmiths may not have been trying to replicate British ironwork as much as they were trying to modify it. It would be best to consider their efforts as the creative adventure of a transitional society. As Britons became Americans, their arts changed. The wrought iron produced during the early national period mirrors those changes. Rather than criticize the simplicity and plainness of these works, they should be read as metaphors of their age—as cautious experiments with novelty.

The nineteenth century brought more changes, particularly after the arrival of three German blacksmiths, J. A. W. Iusti, Christopher Werner, and Frederic Julius Ortmann. Much of the old Charleston wrought iron which survives today is by these men. Indeed, the two most famous gates in the city, St. Michael's cemetery gate (c. 1840, plate 66) and the Sword Gate (c. 1848, plate 67), are by Iusti and Werner respectively. In the antebellum period many important buildings were designed by local architects. Wrought-iron decoration was often designed as well, and a contract for its fabrication sent out for bid. Such was the case with the gates at St. John's Lutheran Church. A. P. Reaves worked up the plan, while Jacob Frederick Roh and his eight helpers built them. Robert Mills is thought to have designed the gates and fence at the First Baptist Church. The City Hall (1801) and the South Carolina

*66   St. Michael's
    graveyard gate.*

67   The Sword Gate,
32 Legare Street.

Society Hall (1804), both designed by Gabriel Manigault, have elaborate stair railings and grills. Nineteenth-century wrought iron is generally ornate, employing numerous scroll forms and floral motifs. The gates at City Hall Park (1824, plate 68) and those at St. Philip's are excellent examples of nineteenth-century decorative trends. Both works have large **S**'s, lyres, and anthemion leaves. This last motif, commonly used in Greek Revival buildings, is emblematic of the period. Its presence proves that Charleston artisans kept abreast of the latest national trends in architecture. Charleston ironwork reflects several social changes which occurred in the early nineteenth century: the melding of diverse ethnic stocks, the elaboration of personal talent and enterprise, and the localization of national styles.

Toward the middle of the nineteenth century, cast-iron ornament became commonplace in Charleston. It was mass-produced, usually in northern cities, and shipped to all parts of the country. Mail-order catalogs featured elaborate drawings and perspective views of intricate grills, benches, gazebos, and the like. Some of these standard items were imported into Charleston, but the John F. Riley Company did some local casting following the accepted patterns. Cast iron grills and gates were much more elaborate than those made with wrought iron. Leaves, stems, and flowers could be fashioned with extreme realism. The intricacy of such ornaments was appropriate for the late-nineteenth-century Victorian modes of architectural adornment, and consequently there was a decline in the demand for decorative wrought iron. Civil War bombardments and an insensitive occupation

68   *Detail of gate, City Hall Park.*

by Northern troops took a toll on the city and caused the destruction of much of the older wrought iron. In the postwar period, wooden fences were often used as replacements. Many of these imitated scroll work with carpenter's joinery (plates 69, 70).

The taste for hand-forged ironwork was, however, firmly implanted in Charleston, and requests for it continued to occupy local blacksmiths. Werner worked until 1870 and Iusti until 1882. Ortmann's sons carried his business on until the 1930s. Both Philip Simmons and James Kidd remember the Ortmanns as the local specialists in ironwork during the first decades of the twentieth century. Kidd recalls:

Ortmann done a lot of this work. He iron off all these part of Charleston. Ortmann done all of Charleston's ironworks, the step rails. I tell you the truth. I think Ortmann pa put them stair rails up on City Hall. Yeah, he done that old-time work a long time. Nobody else would fool with it.

69   Wooden gate, 46 South Battery.

70   *Wooden gate, 44 South Battery.*

As work opportunities for general blacksmiths became fewer during the 1920s, they turned more extensively to decorative iron. Peter Simmons did some repair work on gates late in his career, after he was eighty years old. James Kidd, primarily a farrier, made fences and stair rails when he sensed that "the game was running out." Philip also, as we have seen, turned to decorative wrought iron and thus fully entered into a tradition that began in the 1730s.

Most of Philip's designs derive directly from local Charleston precedents—for example, his gates at St. Andrew's and the First Baptist Church. Indeed, in every case at least one motif or decorative unit can be traced to earlier works. Philip wants his work to look old, so he selects old patterns as models. The various kinds of scrolls, the spears, the fleurs-de-lis, the iron rings, the leaves, the wiggletails, and the flowers are all elements repeated from eighteenth- and nineteenth-century examples. The lunette panels found at 100 Tradd Street and in the Snake Gate repeat the lunette shape of St. Philip's cemetery gate. The overthrow at 67 Broad Street, with its interlinked G scrolls, is similar to an arrangement in a nineteenth-century window grill on East Bay Street. The ventilator grills at the same address feature the broken scroll design used in the gate at the Miles Brewton House, built in the 1760s. The accidental turns of history have placed Philip at the end of a long line of decorative ironworkers, but inheritance alone has only provided a potential role to fill. Philip's deliberate efforts to learn the old-fashioned forms and execute them well have made him the custodian of the traditions for Charleston ironwork, a position acquired by intention, not accident. In this role he is both historian and artisan, for he tells his customers what designs are appropriate as well as providing the finished product.

Philip's wrought-iron sculptures—the snake, bird, and fish—are his most unusual additions to the local ironwork. They are without precedent and stem from his active imagination. But many of the other gates and grills are also unique, even though their content is not new. Creativity is possible within the limits of a tradition, even the rigorously conservative one found in Charleston. New compositions of the usual motifs bring novelty to the 250-year-old tradition. With seemingly standardized fences, balconies, railings, and gates, Philip expresses many of his own ideas about what he considers impressive, "fancy," and "nice." During his forty years of ornamental ironwork he has developed a well-reasoned aesthetic sense which not only directs his hands but also determines to some degree the communal taste in ironwork. At the outset of his career he took orders; now he gives them.

They say, "My neighbor got a porch over there and they got wrought iron all over it, but I don't want it like that." And we sit down and we talk and I come up with something. That's what I like about the blacksmith trade; doing it from scratch and arriving the idea from my own mind, my own thoughts. That's the part I enjoy.

In the midst of his time-bound trade and even while preserving the historic appearance of Charleston, he remains his own man. He has found self-expression in the communal tradition.

4

"So Many Variables Once You Done Get Your Opening"

The Nature of Design

Traditions consist of shared patterns of expectation and performance, and within a traditional occupation like blacksmithing we usually expect that the mark of an individual's talent will be hidden by his obedience to historical precedent. Yet we find that a number of Charleston's historic ironworks have known designers, and that the interesting aspects of these works were usually their unique features rather than formulaic repetitions of standard motifs. Philip often intends for his designs to look old, to resemble the decorative works of nineteenth-century architects and artisans. But a description of the old patterns would not give a full account of his gates and grills. Even in his replication of traditional Charleston patterns he displays a distinctive imagination. Thus a blend of the past and present emerges in his repertoire of wrought-iron designs, a subtle negotiation between the stimuli of tradition and individual talent. When the uniqueness of his Bird Gate (plate 59) was pointed out to him, he responded:

That's what ironwork is all about. When the customer come to you, he don't want what the neighbor got next door. . . . The early craftsmen in Charleston, they took a lot of pride in their work and they, most [people] could tell the early craftsmen works but they look alike . . . but today people want their own thing. Now if they come and tell me they want something like the Sword Gate or like that or the other thing, I'll say, "I don't want to make nothing like the Sword Gate. I could make the Sword Gate but maybe I'll put something in the sword place or something else." That's the pride in the builder. I like doing my own work.

Philip is fully aware of his tradition, but unless a gate uses a novel shape or motif, it isn't very interesting to him. However, once it bears the imprint of his personal touch, the ironwork becomes a source of pride and a reason for continuing in the blacksmith trade.

Designing decorative patterns for ironwork is challenging. The blacksmith has to satisfy his customer as well as himself. A commission becomes more important to Philip if he is free to develop it according to his own plan.

I enjoy doing the work, but they [the customers] come with their own plan. Say, "I want it like this." I go ahead and do it but I enjoy the one when they come and say, "I don't know what I want."

Philip walks a fine line between his own ideas and the requests of his customers. Even though his clients will often take his advice, developing an interesting combination of scrolls and other motifs may still be problematic. Regardless of his customers expectations, Philip carries in his mind the burden of the past. Two centuries of

Charleston ironwork provide the standard that he inevitably will have to use as the measure of his own excellence. When making a fire tool he commented:

It's a pattern of the early century. Well, I deviate from that since I get on my own and I feel very confident of myself, and I deviate from that. Still, it's always something to keep in mind, about the old craftsmen and the contribution the old craftsmen made to this country. I keep that in mind. If I making something now, I trying to make something that would, say, be used a hundred, two hundred years ago.

In like manner, his decorative works must also face the two-hundred-year test, the test of traditional fitness.

I know how long wrought iron supposed to last. I build a gate, I build it to last two hundred years. If it looks good, you feel good. I build a gate and I just be thinking about two hundred years. If you don't you're not an honest craftsman.

Because Philip, above all, wants to "do his own work," he constantly takes the more difficult approach to a design problem, the search for a new order. Ideas are constantly racing through his mind, but his mental circuits run a feverish pace when a customer has a job to be done.

Guy said, "Here, this what I want, Mr. Simmons. I got a staircase. It's inside, this is not outside. I want spokes but I don't want it all spokes. Want two sets of decoration. Suppose you put one here, Mr. Simmons, and one up here. Then you can go ahead with your spokes. Leave that opening in there. Now put something fancy on there. What could you put in there for me?" I say, "O.K., let me think, think of something." All my ideas gone.

But what I would do was put this on paper, then I'll take it outside and I'll put several [designs] in there. I go 'head to my work and don't just try to bump my head to tell him what I can do. Don't think about [it], just go ahead. I come in the evenin'. I'll lay down in the bed. Visualize it. Then [at first] things divided in your mind. Then you don't see too many [options]. But I come up with something. . . . I had to stop and think. Have to wait. I have to clear up, what you call it, the air traffic. When the plane come in and can't land and he say, "Fly around." Then he'll tell the passenger on the plane, "We're waitin' till the air traffic clear." Well you have to wait till that traffic clear up on my mind.

Over the years Philip has learned that the plan of a new design will eventually come to him. He does not develop it outright; he waits for it to arrive.

It's like a Boy Scout, you got to always be alert. Sometime you go ahead on a piece [of] work, doing it. You don't know how it going to end up. Day by day you're thinking. Go home, you lay down in your bed. I got a [note] book here that I use. I sit on my bed and I carry it to the bed and I lay down and I think and I read over and I make a little notation. And I had that little book with me. I sleep the whole night and I'll get up and I'll make a little notation. And I come in here [the shop] and put it together.

Sources of Design

Philip claims that he "arrives" designs from his mind, what other black folk artists sometimes refer to as "getting futures." In either case, reference is made to a dreamlike vision bordering on prophecy. When asked if his designs came in dreams, Philip answered:

No, I don't dream about it, but I have a vision. I have a tremendous vision. Every so often it comes over me. I never dream about it. I get a tremendous vision every so often. I'll visualize something. I never forget about it. I see it finished completely in my mind.

At this point Philip sketches what he is able to picture mentally and then translates his sketch into iron. Even though Philip explains the sources of his inspiration in instinctual, even mystical terms, his creativity is far from automatic. With more than half a century of wrought-iron experience behind him, it is more likely that forming a mental picture of a design involves making a long inventory of previous works, a review of potential assemblages, and hypothetical tests of new motifs. Once all the options are sorted, "cleared up" in Philip's terms, the form most appropriate for its context emerges as the best solution. Since Philip deliberately works within the boundaries of his tradition, the choices he makes tend not to be immediate, but tested. Thus the final design "arrives" at last rather than being invented at once.

One design that Philip has in mind has yet to reach the point of maturity. He intends someday to make a complex piece that will capture the history of the city of Charleston. Since this vision is only partly realized, his account of it gives us not only some insight into the content of his creations, but a sense of the creative process.

I haven't worked it out yet, but what I had in mind by living on a peninsula, I be thinking on a lot of outlying farmers and fishermens and stuff like that. Like the fleet landing, there's the "Mosquito Fleet" in Charleston. They used to go out and make their living in small boats. My mind is just crowded and I hadn't worked it out yet. I want to make something that people could just come and look at it, and say, "I'm on the peninsula. I'm looking at something made on the peninsula." Or something to carry your mind back like a fish or something. Or some type of fruit or vegetable 'cause Charleston area is . . . not a fruit, not so much a fruit around Charleston . . . a vegetable or a fish or something. Something of the river or farm. We got a lot of seamsters around here, maybe about . . . I don't know. It'll be a boat or fish or. . . . [What about Sea Island baskets?] I had that in mind too. Just hadn't work it out yet. But my mind is crowded with so many things that was made around Charleston in the early century.

I don't think it would be a map. It'd be something somebody could walk by and read and get ideas from. It would have a lot of illustration. Yes, I had several things in mind . . . like, say for instance, the Sword Gate that represent the old fort, the war, the War with the States then. Now I'm thinking about so many different trees around here, so many fish in the river, and I'm thinking about making something to represent the fish in the sea, the tree in the woods, a flower in the garden or whatsever. I'm thinking seriously. I haven't come to a conclusion or decided what I would be making, but I look at nature every day and several things arrive from it. Just flash over and flash over.

This design, what we might title the "Story of Charleston," is developing gradually. Philip has been considering it for more than three years, but other commitments take precedence because there is no client for this gate, grill, fence, or whatever it turns out to be. But obviously the kernel of an ideal plan is present in his mind. Referring to a vision he received concerning this emerging ironwork, Philip notes:

It come in the form of something with a historical background. Every so often something different will appear because in the peninsula there are so many things with a historical background.

This new design is thus set in the past and will be affected by the local tradition in intent, if not form. In the meanwhile, Philip continues to take inventory of appealing motifs which he pulls out of his experiences. Some motifs, the fish and vegetable, recall his Daniel Island boyhood, old wagon wheels bring back the memory of his apprenticeship, and the Sword Gate symbolizes antebellum Charleston and the decorative ironwork tradition. Because this design will eventually be created from the diverse experiences of his life, its process of composition is very contemplative. Only when he has a full sense of his own historical contribution, his significance to the community of Charleston, will the design finally be completed.

Every so often now I have a vision that I think some day will materialize. I don't think it will be too long now. At my age I'll have to do something pretty soon.

Philip often traces the sources of his designs to natural forms, particularly trees and their leaves.

Maybe I can arrive something from my mind just by nature, looking at nature. Sometimes it might be a tree. We got palm trees around Charleston. You get up in the morning, you look at nature. Then you look at other things around Charleston. Then you stop and think and you say, "Someday I'll make this or the other thing."

While a tree may provide an image to work with, his rendering of a floral motif is hardly simpleminded copying. He abstracts the flower, the tree, or the leaf into an ideal form stamped with his own sense of order. His gates at 5 Stolls Alley (plate 10), the overthrow of the Bird Gate (plate 59), and his magnolia leaves in the stair rail at 107 Tradd (plate 57) all provide examples of nature rendered in iron. The pecan leaves which Philip now often uses as a secondary decorative element are a particularly keen instance of creativity drawn from a natural image. Each leaf is unique, with its own dimensions, shape, and pattern of hammer marks, although to the casual observer they would all seem to be the same. A section of iron rod is doubled back upon itself, mashed flat, and drawn out to a pointed tip (figure 1). The narrow open split represents the large central vein, and the random hammer marks simulate the smaller veins of

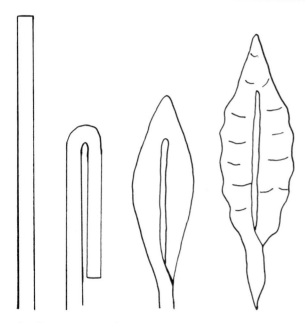

1   Steps required in the making of a pecan leaf.

the leaf. These features Philip verifies by yanking a low-hanging leaf from one of the large trees in his yard and holding it up for comparison. He goes on then to explain that "no two leaves look alike. They are similar, but they are not identical. That's the way I make mine. No two alike." He may even, in jest, offer a challenge to the onlooker. Pointing to a towering tree laden with leaves, he will offer a prize of five hundred dollars if any two leaves can be found which are identical: "I guarantee you won't find any and if you do it will take you so long and you have to work so hard. You'll earn that money and have a long gray beard too."

Philip's observation of the variation in nature leads then not only to imitation, but to improvisation. He can make each leaf exactly

as he desires at a given moment. He can make it curve slightly to the left or right, he can make it smooth or crinkly; its tip may be sharp or slightly blunted. The possible variations are endless and yet, paradoxically, while following the random impulses of his personal inspiration, he is simultaneously pursuing an image established in the natural realm. Formal precedent and individual imagination are combined.

The Process of Design

The general shape of a work of wrought iron will usually suggest what decorative motifs may be used. The structural frame of a gate or grill establishes the limits of shape and dimension. Thus a small grill will very likely require a series of small, tightly curved scrolls to fill it up; large motifs would not be appropriate. In the design of decorative works the pragmatic aspects of the blacksmith's craft must always be considered. The gate must swing between the pillars in the wall, the grill must fit the window opening, the banister must conform to the rise of the stairs. A decorative piece of wrought iron must serve its purpose well or it will never have much aesthetic appeal, no matter how beautifully it might be rendered.

The first step in designing a commission is, then, to study the shape and size of the required ironwork. This Philip explains with reference to the railings on his own porch (plate 71).

As simple as that thing [is], that star there, those spear there. I didn't want no scroll. I

71   *Detail of panel on Philip Simmon's porch railing, 30½ Blake Street.*

said I want it [ironwork] on my porch, but I didn't want no scroll. I think about what Sears was puttin' out there. You want to get away from it. Okay. I lay down in the bed and say, "Why I can't put some spears in it? Well, what you going to put in the column? Just make the column simple." Here I come up with one bar. Easy to make, but it something different. So like that, you can always think at leisure. Things don't crowd your mind. Get the outline 'cause you know you've got to have that.

Philip's porch rail is very plain. It is so austere that one might not give it a second look. For Philip, however, it is significantly creative (although he wouldn't call it fancy) since there is not another railing like it. There are no curved pieces, only straight square bars. Beyond this it was "easy to make," thus an efficient and practical solution to the problem of providing a support for the roof and a guard rail for the porch. On Philip's house the roles of craftsman and artist were both performed; a function was served and a unique design, albeit a simple one, was created. This blend of what is usually considered two distinct realms presents no quandary for Philip. When asked to distinguish between art and craft, he responded:

I don't see no difference. That's the way I see it. When I say I'm an artist in my own area [blacksmithing], I was referring to the decorative part when I finish making it. You're looking at two different phases when you thinking about hand work, you know: craftsman and artist. Well I do the decorative part and the fabricating part too.

Philip's recent remodeling of a banister in the Westmoreland house on Tradd Street provides a good example of this duality. The Westmorelands wanted the height of the rail raised seven inches. Philip noted that the space at the bottom would have to be blocked off by a flat bar. He also advised that some further decorative elements be added.

I suggest to put little rings underneath there 'cause he just said put the bar. The bar would look kind of open and if you step on it—a kid might get on it—it'll be flexible. Instead of puttin' just another short spacer in there, I just put the wheel [ring]. You got your support and you got the decorative part. I mean the decorative part is the wheel. So then a kid could stand up on it and it won't bend. Yet they got something nice lookin' to look at. I had want it to look good.

Techniques of smithery learned from the Old Man emerge in the decorative work. Comparing horseshoes to porch rails, Philip observed:

It's the same bending. Heating and bending. It looks different, but it the same principle. When you heat iron and bend it the shape of a horse hoof and heat iron and bend it in the shape of a porch rail, you're making some kind of circle. So it's the same principle, but you've got to adjust yourself to both.

Philip's works, as we have seen, are certainly well crafted. Beyond this they carry special qualities, having been first imagined and visualized as unique ideas which are then carried into action. Philip calls himself a craftsman, for that is the title traditionally given to blacksmiths. Trained first as a tradesman, he has never abandoned that image, even though his current reputation has been built upon artistic commissions. He is quick, however, to point out that once something is made he then "puts the decorative on it."

His principles of design intertwine the past and present; the way in which these principles are set in motion is similarly complex. Presented with a hypothetical design problem, Philip quickly, without a moment's hesitation, proceeded to explain just what he would do.

Well, if I'm going to do the measuring out myself, I'm going to allow myself enough [room] to get a center in there. But if I'm limited to length or height or depth then I'll have to find my center. The center's important in some of my work. Sometime you cannot get a center. So you have to use three-thirds of your iron instead of one-half. Split [the middle third] and put your center in there. Sometime it's not balanced out. So you have to divide it into three parts instead of one, in order to get the center. Sometime cannot just find one center, I'll have to get two. Take a rail like that with seven spokes. Then you know the center spoke must be the center one, you get three spokes on each side. Suppose it isn't seven bars and that thing is eight bars. You see then I've got to find a center. Then I got to get away from the center bar and use two bars for the center.

Suppose somebody bring a piece of work to me. Tell me, say, "This what I want," and I can't find a center. Then this the part I got to create, the decorative part. Because that is the

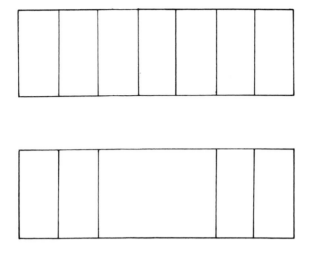

2   Steps in the design of decoration for an eight-spoke railing.

decorative part in there, that center piece. Sometimes it's impossible to find a center without being odd. Let me show you what I want, why sometimes it isn't easy to find a center. [*Philip begins to draw a sketch.*] All right, here is one spoke, two spoke, three spokes, four spokes, five spokes, six spokes, seven, eight. Now where would you say the center is? This the part . . . in doing the decorative part this is where I got to start. Now where is your center? Right here isn't it? [*He points between the fourth and fifth*

spokes]. Okay, now what would you do here? You done find your center for me. Here's your center here. Now I got to put some kind of decorative in there. So then I come up with this sometime. Then I take these two [spokes] out and you got the same thing (figure 2). Now then I join this [curve] here. See you got a wide space. Somebody could go through that, a kid could come through there. This the decorative part. Now you could set **S**'s just like I first started, but I got to use that spoke there. Shortened [the **S**]. Use that one, stop up in there. But he look better by takin' out two spokes instead of just using this one spoke here. You got to use those two [spaces]. Then you got to fill it back in.

Qualities in Designing

Symmetry

Since Philip's designs follow the patterns most commonplace in Charleston, they are almost always symmetrical compositions. (Only his animal figures do not conform to this rule.) Thus the location of the center line is crucial. Once determined, the various motifs fall into place on either side of that axis and, in Philip's view, creativity begins.

You think about the middle. In all my work I think about the middle. Very easy when you think about the middle.

Shape

The shape of a work, its outline, automatically conveys both a sense of genre and an appropriate decorative motif. A square space suggests a window grill, a tall rectangle a walkway gate, a horizontal rectangle a driveway gate. Once the nature of the working space is determined, decorative ideas "arrive": "If you don't see an empty space, then it don't mean nothing to you." Discussing a window guard measuring two feet by four feet, Philip elaborates:

You say, "Well, I want a nine-inch scroll." Then you say, "That's too open. Why don't you put two scroll four and one-half inches." Then you can start putting it on paper, after visualizing it. I call that the move to fill in. Now you got the major thing [decorative element]. If you don't want it that large then you cut your size down. Now how 'bout a scroll to fill this up. See this [space] what you got to go by. Now you put that [motif] there. Now let's look at your opening now, see what you got. Now this is open. So what you going to put here? If you wanted, you draw that [a curve] right on down and put just that there.

3   Philip Simmons's sketch of scroll work.

Now you done fill up. If you're too open there, all you got to do is put a branch like a tree in there. Why don't you do something like this [a reverse **J** scroll]? Right here is open. Why you don't put a bar in there and top that and put spear right there? You can keep thinking 'cause you got this [outline]. You just keep thinking now. You can extend this spear point. Don't let it touch, come close there . . . there are so many variables once you done get your opening (figure 3).

Density of Ornament

Since the decorative elements in a piece of ornamental wrought iron are often placed in one spot, their arrangement will be watched closely. Philip describes ornamental units either as open or clustered, and a good design strikes a balance between these two qualities. Referring to the Sword Gate, he noted:

I wouldn't put so much [decoration] in one place. I'll scatter it out. It look all right, but if I was going to make it I wouldn't make it like that. I rather put one or two **C**'s in it and let it stand out. Put scrolls most around so it'll look attractive.

The visual balance between a densely clustered design and one which is spread out enhances two aesthetic qualities: visibility and clarity. When scrolls and leaves are positioned with some degree of relative isolation, they stand out and can be seen more clearly. Philip wants each of his efforts to be appreciated even though the whole design is more important. Thus he struggles to "get it right" whether drawing the design on paper or fitting the many parts together.

Curvature

Scroll-work elements are the most common decorative units used in Philip's fancy pieces. They usually resemble rounded letters: **S, C, J, U, G**. When these shapes are turned, Philip is careful to make sure that they are perfectly curved. He has no exact term for this ideal quality, but he is always certain when it has not been achieved. In checking over his apprentices' work, one of the first flaws he will point out is likely to be the execution of the scroll work. He will instruct his helpers "to make a circle" with an **S** scroll; that is, to reshape its ends until they are more rounded. Philip regards the older ironworks of Charleston as good examples of excellent workmanship because not only are the scrolls nicely curled, but the curls spiral gradually to a small coiled circle at the center which he calls an "eye." These are features he repeats in his own work in order to "show pride" in his craft. It is no accident that Philip feels a kind of professional kinship with the Afro-American makers of coiled-grass baskets. He points out that "they curl around so beautifully." Basketsewers have a variety of shapes, but Philip focuses on those which resemble his scroll work, and thus he reveals the importance of appropriate curvature as a criterion for the evaluation of ironwork.

Other shapes with curves in them, such as the openwork spear points that often sit atop gates, are also checked for their rounded lines. The spear should be symmetrical and come to a dramatic point at its tip. If it is too angular Philip will note that it needs to be "puffier," yet if it curves too much he may criticize the form for "too much belly." His notion of correct curvature is obviously an ideal arrived at through a balancing of options. Like the density of ornament, the curvature of wrought-iron decoration is a relative quality.

Precision

Philip often thinks about his designs in terms of dimensions; he has to, since all of his works are tailored to fit a specific space in a specific house or yard. But more than this, he studies the dimensions of the materials he uses, understanding that even a simple straight spoke in the plainest of stair rails will have a visual impact. Depending upon the length required and the intervals of the spaces between the bars, he may select ½-by-½-inch or $9/16$-by-$9/16$-inch. The layman may not consider a sixteenth of an inch very significant, but such dimensions affect the sense of mass of a stair rail and can make it appear visually full and dense, conveying a feeling of strength, durability, and security. Philip understands from long experience how the sizes of materials must be matched to context and function. In a walkway gate he does not use bars of the same size as he would usually place in a window guard, even though similar motifs may be used to decorate both objects. This differentiation is determined by a sense of proportion similar to the sense of visual balance used in ornamental arrangement, except that here the aesthetic decision is expressed in finite measurements.

Over the years Philip has refined his ability to judge distance to such a degree that he knows the exact length of a line, the span between pillars, the diameter of a bolt, or the size of a piece of stock, just by looking at it. When he made the fish in the Star and Fish Gate he wanted it to be twelve inches long. He

*72   Making the decorations for the Star
and Fish Gate.*

4   Rendering of the Star and Fish Gate.

took no measurements but proceeded directly
to cut the required slats and bend them into
the outline of a bass. "I was conscious of every
piece and how much I want in there and the
size of the fish" (plate 72). He centered it in
the midst of eight spokes in the lower section
of the gate. Almost two years later, when
shown an attractive pen-and-ink rendering
(figure 4) of that gate, he immediately pointed
out that the fish was drawn incorrectly. His
fish, he noted, was longer, and it extended well
past the third spoke so that the eye was to the
left of it, not to the right. Philip remembers

precisely what he has done because his design
process depends so much on exact and precise
measurements. Rigorous discipline in
handling the materials gives the artist a
strong feeling of control when making
decisions in which there are several choices
that are available.

Philip's ability to gauge distance by sight
alone was also shown to me on another
occasion. After he had finished making an
old-time log roller complete with a fancy "rich
man's handle," I asked him about other styles
for this tool. He responded that there were

other ways to make log rollers, and, taking up his chalk, he sketched another form on the floor of the shop. It was very much like the one he had just made except that the handle was a simple ring, a "poor man's handle." For no reason that I can remember, I placed the tool he had made down on top of the chalk sketch and found that not only was it the same length from pointed tip to the end of the handle, but the hook was located at exactly the same spot and had the same degree of curvature. It was then evident that Philip possessed an uncanny ability to judge distance and proportion. When he made something, he was able to compute to the fraction of an inch how it should appear. He then directed his hands, his tools, and his materials until the proper shape was formed. "Eyeballing" was not guesswork for him but a deliberate judgement verifiable not by the ruler, but by fifty-five years of experience—what we might call a shopwork tradition.

Improvisational Forms

Because the right balance between openness and closedness only emerges from the process of design and fabrication, Philip's creation of decorative wrought-iron works is essentially an example of experimental composition. He has the outline of his work which he then fills with ornament. The "move to fill in" begins with an almost tedious freehand drawing, which after years of experience closely follows the scale of his measurements—usually one inch represents one foot. Scrolls and leaves are sketched out in pencil, evaluated, erased, and redrawn—often in the same place. This stage of design may last a few hours or a few days, depending upon the complexity of the commission and the urgency of the job. The

sketch becomes more and more ragged, but despite its final tattered appearance will be consulted closely and often during the fabricating process.

Once the outline of the piece is laid out in iron bars, the "move to fill in" is repeated, although this time it is with chalk lines on the floor of the shop. This second drawing is a full-size, scaled-up version of the original sketch (plate 73). This is similar to the boat builder's process of lofting, wherein the lines of a plan are scaled up to check the fine points of design, which may not have been noticeable in the original smaller model. Again the drawing is done by eye, but with exacting precision. A scroll, for example, will be drawn several times until its curves have not only the proper degree of roundness, but are also coiled into enough turns to fill up the space visually. Measurements for the final wrought-iron elements are taken from these lines, but once made they are not always set atop the chalk-line template. Instead, the design is reevaluated in a third round of manipulations. The various parts are shifted about and other possible arrangements considered; secondary improvisations enter into the design. Philip believes that his open-ended approach improves his work by providing him with the option to incorporate changes before too much of the design is set and thus difficult to modify.

It isn't always a thing gonna be set in your mind and when you just half way you can see you ain't gonna like it. Sometime I draw the whole thing and don't like it myself, not the customer. What that comes from . . . you think you like it to start . . . it isn't always you like something you can visualize. But one thing, you can visualize, it give you a

73 Philip Simmons's sketches for a decorative commission.

background like this thing [drawing] here. I may not like these scroll when I start, but still I see it that way after puttin' it in and I see where I can improve it.

Just as Philip's ideas of design emerge, gradually becoming clear visions, so too do his pieces themselves gradually take on their final form and order. His approach to the artifact is clearly improvisatory or innovative. It is not just an expedient technique for making complex objects that he follows, for often the net result of his three stages of manipulation will be a piece which is exactly like his original drawing. But instead of approaching that goal directly and following the imperatives of the established plan, he comes to the final product by a roundabout route. Philip gains insights into the nature of a piece from making it as well as thinking about it. The drawings he makes guide his eyes and hands, but the hand work may in turn inform his eyes what to see.

Vitality

In two of his works which employ animal sculptures, Philip took care to develop a sense of being or presence of the represented animal. The snake (plate 36), for example, was "dead" until the eye was properly positioned. Even though its body was shaped into an undulating triple-reverse curve and it had a full set of rattles, a threatening angular head, and forked tongue, the snake lacked the appropriate spirit until its eye was correctly placed. The correct position was, according to Philip, a precise spot in the head that made the snake seem to look back at the viewer. However one moves in front of the gate, the eye of the snake, he claims, will follow. This roving gaze, then, is the most important feature of the sculpture, and it was definitely the artist's imaginative vision that brought out this quality. The snake form was requested by Philip's client; it did not come in one of his visions. As he put it: "That was a vision of somebody's else. That was already in the making 'cause that snake represent the Gadsden house and the owner wanted something that speaks 'Don't Tread on Me.' And that where the snake come up." Yet in his effort to render the snake, he did impose something of his own vision or sensitivity onto the raw metal. Making the snake come alive, giving it a dynamic presence, was more important than the virtuosity of execution, the appropriateness of motif, or the beauty of design. Philip gave the snake more than form. He imbued it with the essence of a serpent. Philip admits that making a snake was not his original idea, but he quickly adds: "I had to make it come alive. That's the important thing."

The fish (plate 64) in the Fish and Star Gate was fashioned with a single species in mind, the striped bass known locally as a "spot-tail." This type of fish is marked by five dark lines running along its sides from head to tail. Philip, who once fished as a boy with his grandfather for a living and who still fishes as often as he can, knows this animal well. This familiarity led him to use five slats, each marking one of the five lines in the fish's body. While he notes that this arrangement of pieces creates an open effect, it also recreates the distinguishing feature of this particular fish.

But Philip tried to make his sculpture realistic with more than markings. He made the mouth open in order to suggest a fish in action. This particular bass was feeding: "he mouth open 'cause he goin' after a shrimp." Indeed, as the sculpture was positioned in the lower section of the gate, Philip developed an involved scenario about his fish. First it was angled with the head tilted up and then pointed down. Philip spun a tale: "The fish is swimming. He come up for the bait, then run down from the boat." The fish was eventually left in its final horizontal attitude, but Philip continued to comment, "If you have the right bait you can catch 'im." This statement was meant as a humorous lure to his audience and drew the intended question, "What bait?," to which Philip responded, "Cash money." But joking aside, the fish still possessed a quality of vitality, of being alive, like the snake. The eye was again important, and Philip said that he would have liked to put a marble in the eye hole to make the sculpture look more realistic.

Even the star portion of the gate was given special attention so that it might seem more lifelike. Two five-pointed stars, one inside the other, are set in an iron circle. Philip calls it the "star within a star," and noted further that the motif is "like watchin' a star, it get smaller on you." Thus a sense of twinkling movement is captured, movement which signals the presence of energy. Iron is obviously stiff and unyielding, but in Philip's mind this stubborn medium can be given shapes which will capture the idea of motion and thus be transformed into vital and interesting designs.

Finish

Once a piece of ornamental ironwork is completely assembled, further aesthetic qualities regarding finish are considered. Some clients regard blacksmith work as rough and rugged. They want to see evidence of handwork and hence ask that the dents of the hammer blows show up on the forged surfaces. Philip, however, prefers a smooth finish which demonstrates more control over the techniques of forging. He also carries this notion of smooth surface over to works in which the various elements are fastened with his electric arc welder. In such cases the melted welding rod leaves an irregular lump of metal which fuses the pieces to each other. These lumps are removed with a hand-held grinder so that only the bits of melted welding rod that fall into the thin space between pieces remain. After grinding, a weld will appear pitted. It is then smeared with a glazing compound which fills in any imperfection, a process Philip refers to as "fulling in the cracks." In the end all the parts show few signs of their connection; the object appears as a single complex whole. The unity of finish throughout contributes greatly to this feeling, and consequently surface quality is monitored closely as a piece nears completion.

That's is important in all work, that smoothness. When you dealing with smoothness that's applies to *all* work whether it's round when you grind it off, [all] got to be round. If it's square, it got to be square like the original iron. If it's flat, he got to be flat like the original one. Regardless of what kind of material you're dealing with, what shape, it got to be smooth. You notice some of the work apprentice doing around here, you can see places where it should be smooth, it's not smooth. You don't supposed to see no weld on no work, 'though it's weld.

Color

In order to protect the iron from rusting in the heavy salt air of Charleston, Philip normally gives his finished works a coat of "red lead" or anticorrosive paint. The final color is usually left up to the client, who in most cases will choose black. Dark colors have been prevalent in Charleston ironwork for the last hundred years—either Charleston green (a dark olive color) or black. It is interesting that the earlier ironwork in the city is described in some accounts as light blue or yellow, pastel shades which brighten up the ironwork and the buildings to which they are attached.

Philip prefers lighter colors, not because they represent the oldest set of color choices in the community, but because the detail in the work becomes more clearly visible. This is consistent with his notion of open, uncluttered design. Each element should be shown off as a distinct form and should then stand out from its background. Dark fences and gates are often obscured by a background of green lawns, hedges, and shrubbery, whereas a white fence can be seen clearly. Dark paint can hide imperfect welds and rough surfaces, flaws that Philip's works do not have. It is no wonder, then, that he would rather have his smoothly finished ironwork coated with white or light blue paint instead of Charleston green. "I like to paint 'em a light color so you can see the detail. This black is too much. Hides the work. You get close and you still can't see what it is." He may warn those who ask his advice on the choice of a color: "If you got a white house and you paint the spokes on the window guards black, then it gonna look like a jail."

Customer's Influence on Design

Beyond the complexities that Philip imposes on himself, constraints of personal excellence, there are issues of propriety that are imposed by individual clients and the whole Charleston community as well. Some people come into the shop with a firm idea of what they want. They give specific directions or provide a photograph from a book or magazine they have read. Often these potential customers will present cast-iron designs which could never be rendered with the blacksmithing techniques of ornamental wrought iron.

They bring in a piece of cast and ask to make it do what a piece of wrought iron would. Non-possible. They will bring a piece in the shop for I to make things up. I say, "That's cast iron." "Oh, I didn't know."

These people need to be educated, and Philip gives them lessons—subtly. He obviously does not want to lose their business, so he must cautiously transform their original requests.

They'll tell you what to do. You know it's not going to work. They want something just to satisfy their eyes. I would always want them to know that isn't the best thing to do. As well as you come in there with something, I got to play a part in there too. I will always keep that in front of the customer: what your eye see, it isn't what you should have.

A complex design involves more labor, more materials, and more mental effort, and consequently it costs more. This economic factor may then have an impact on the levels of creativity expressed. Referring specifically to window grills, Philip explains:

It's all depends on how you want it to look. And you got to keep the price down. People don't want to pay. This decorative part we put in the center costs more than that [plain grill]. But now we could sell this [simple design] to the customer easier and cheaper and they likes it. That's why we try to stay with so much bars. People can afford that instead of fill up all these with scrolls, like some of it you seen me done. When you start bending up this, you going to run into money. When you're doin' a piece of work or someone come to you, you got to give them something nice lookin', got to give them something with a reasonable price. So you see these quarter **S**'s I got in here, people like that and I could do that cheap.

Usually with the commissions requiring fancy work, economic considerations are secondary. In restoration work clients want recreations of eighteenth- and nineteenth-century designs and are more concerned with historical accuracy and social symbolism than with cost. Philip, in such cases, is called upon to interpret the local tradition in order to enhance the feeling of antiquity of a given building. This is no simple task and is further complicated when a client wants the grandeur of a more eloquent image than history may have actually allowed for his house.

When you go to work up like on the one on Tradd Street, you run into a hundred, hundred fifty dollars [for a window grill]. Those people, that's what they want in that section. Not that he want to pay for it, but that's what the society of that area got and that's what that area wants.

I try to stay with that and I let them know it too. Go to some of them, they don't know what the community wants. I can tell them 'cause I study it, you know. I study every area, what I can put up. That house over there [32 Blake Street], what I put on that house I can't put it below Broad Street or I can't put it nowhere here around Charlotte Street. Those owners of years ago owned two, three houses and you notice they built alike, they paint alike, they designed alike. The decorative on there, that's alike. So you see you've got to stay with that. A lady say, "I want it." I say, "Well you know these three houses, they're sister houses. I don't know whether that'll work in this community." I don't go to the [Historic Charleston] Foundation Society and ask 'em, I know. "Well what you think'll work in here." I just make a sketch. I say, "This'll work good." I say, "Even if it's not what your neighbor got next door, it's something different." They say, "Okay, go ahead and put it up." But I notice in doing this work you got to know the different neighborhood. Some people, you'll go there, don't know what's all about. You got to study this thing. Put a lot of time in studyin' it. Just like you get books on Charleston—study the wrought iron. I walk around here and see what the other man got.

When Philip Simmons finishes a gate he has satisfied himself, his client, and Charleston's sense of history.

Most contemporary artists sense little social restraint upon their works, but in the creation of ornamental ironwork in Charleston the direct influences of the community are

Charleston

The numbers on the map correspond to the plate numbers in the book and indicate the location of Philip Simmons's work.

The letters on the map refer to significant places in Philip Simmons's neighborhood.

A His first house on Vernon Street
B His second house on Washington Street
C The Buist School
D Peter Simmons's shop on Calhoun Street
E Philip Simmons's first shop on Calhoun Street
F His second shop on Alexander Street
G His third shop on Laurens Street
H His current home and shop on Blake Street

As Philip Simmons defines Charleston, Calhoun Street is the dividing line between the uptown and downtown areas, and uptown is divided into east and west sections by King Street.

inescapable. Referring to the borough divisions of the city, Philip says, "You got to stay within that village or what they use in the early century." Philip is not perturbed by the creative constraints imposed by his customers. He sees their requirements only as one added demand in what is already a demanding process. While he notes that the Historic Charleston Foundation sanctions certain uses of materials and forms and says, "You got dance by their music," those guidelines do not prevent him from simultaneously following his own visions. For a folk artist the demands of his community and his personal motives are generally compatible, if not identical. It is important to remember, however, that the decision to repeat the old, the commonplace, the typical, the usual, can be as significant as an innovative choice. Tradition is kept alive by a series of deliberate, rational decisions. It does not survive without talented, committed performers. Repetition can thus be seen as creative, even heroic, rather than stultifying. Indeed, we have seen in Philip's own discussions of the creative process how much improvisation can be allowed within the context of a community inspired by its history. Clients come to Philip's shop with full confidence that whatever he makes will not only be of high quality, but appropriate for Charleston.

I asked them, I say "You got it in mind of anything you want?" "No, I want you to design it yourself. And just what you make, I'll know I'll like it." I go ahead.

By giving his customers the designs they want, the forms they expect, Philip keeps his work traditional. Nevertheless, since his work is done in an inventive manner and because he maintains strict standards of quality, he has earned the esteem of the community as a master of iron, an individual and personal honor. When one of the other local ironworkers calls Philip the "dean of Charleston blacksmiths," the title refers not only to his age but to his excellence.

Ethnic Heritage in Charleston Ironwork

Although we gain important critical insights into Philip's work when we understand his personal viewpoint and the attitude of his audience, there is a major cultural influence still to consider: his ethnic heritage. It is unfortunate that so little has been written about the lives and work of other Afro-American blacksmiths. Until we have more material for comparison, we will not be able to say how typical Philip's approach is for black American ironworkers. Yet a number of Afro-American artists and artisans have been studied and their comments on creativity recorded. When this body of criticism is compared to Philip's statements it becomes clear that his perspective is not wholly unique or idiosyncratic. Rather, his expressions in iron are but another manifestation of an Afro-American approach to the artifact, an alternative view of the world and the meaning of objects.

Many black creators claim to receive their inspirations in a spiritual, almost supernatural manner. Some, like Sister Gertrude Morgan of New Orleans and James Hampton of Washington, D.C., were visited by God in visions. Sister Morgan was called to preach the Gospel, which she did on street corners, but later she received a command to stay indoors. At that point in her life she turned to drawing and painting her visions of biblical scenes and other inspirations, such as Jesus and her flying over New Orleans in an airplane. Hampton, who died in 1964, left behind an elaborate assemblage of decorated furniture called "The Throne of the Third Heaven of the Nations Millenium General Assembly." Covered with shiny gold and silver foil, this work was his response to a series of visions beginning in 1939. Some of them are recorded on tags attached to various throne elements: "This is true that the Great Moses the giver of the tenth Commandment appeared in Washington, D.C., April 11, 1931."

Harmon Young, a Georgia wood sculptor, talks of "futures" which come in dreams and provide the basis for his carvings. Similarly, Leon Rucker of Lorman, Mississippi, credited one of his highly decorated walking sticks to mysterious visionary sources: "The idea came with the voice of the man. Now who was the man, I don't know, but I say he must been a god 'cause men couldn't do a thing like that just by himself." Other black artists point to the mental aspects of their creative insights. Clementine Hunter of Melrose Plantation, Louisiana, says, "I just paint by heart," but also points out that ideas do not come automatically: "To paint a picture you have to study your head." James Thomas, a clay sculptor and bluesman, agrees: "If you ain't got it in your head, you can't do it in your hand." Thomas's approach to sculpture is astonishingly similar to Philip's, even in the words chosen to describe the process: "The dreams just come to me. If I'm working with clay, you have that on your mind when you lay down. You dream some. Then you get up and try."

Thus, when Philip speaks of receiving "visions," they are more than individual inspirations. His use of a supernatural image is in fact a perpetuation of a metaphor in widespread use in Afro-American culture. Black artisans see themselves as moved by exterior forces outside of their control. Those who are extremely religious, like Sister Morgan and James Hampton, end up making religious art, but even those who are more secularly oriented hear voices, have images revealed to them, or have forms arrive after "deep study." By receiving visions, Philip is not only united to other Afro-American artists, but to traditional African artists as well. West African mask carvers often claim to dream the mask before they sculpt it. Among the Gola of Liberia, for example, it is believed that an artist's spiritual guardian or *jina* provides him with the genius (*neme*) to create a good mask or statue. The Gola people often characterize their carvers as dreamers. The manner in which Philip receives inspiration, his conceptual process, allows us to consider his works as more than an element of Charleston's ironworking tradition. Viewed as products of "visions," we may see them simultaneously as an achievement set in the realm of Afro-American art.

A second feature of Philip's work which can be used to link him to his ethnic community is his use of improvisatory techniques for both composing and fabricating his ornamental ironworks. Whenever Afro-American creativity has been discussed at length, a blend of broad general structure and highly variable content has been noted. Among Bahamian narrators, for example, the concept of "old story" remains firmly fixed despite deliberate attempts to shuffle and reorganize set "motif clusters." The old story has latent within it a vague sense of structure—opening, event, closing—which promotes innovative modifications with stock elements. The chanted sermon encountered in black churches, in similar fashion, has a general outline which is punctuated at given intervals with formulaic epithets or spontaneous shouts and praises. The contests of insult, "doing the dozens," played by blacks on urban street corners, involve an exchange of rhymed couplets. This pattern is fixed by the rules of form, but the impact of the game stems from graphic emotional images set into the stanzaic formula.

Afro-American music has long been acknowledged as improvisatory in nature. While blues songs commonly employ a twelve-bar structure with specific chord changes, the bar pattern can be shortened or expanded. But more important, a wide range of tonal blends, twists, slides, and dips can be interwoven with the standard chord progression, giving each performance a unique feeling. In addition, the rhythmic bass line so dominant in black music contributes an ever-present counterpoint to the melody of the lyric, making even the simplest of songs complex and innovative.

Given the preponderance of improvisatory activity in speech and music in black culture, it should be expected that black craftsmen would also employ an improvisatory approach. We see in Philip's work a reliance on experimentation and spontaneous changes. When we observe him developing insight about an object (the "right feeling") while making it, we can profitably compare him to a jazz musician who explores the potential impact of various sounds in the midst of performance. Philip has an approach to his art works that can be described as Afro-American. By working in an improvisatory manner with dream-inspired designs his ornamental wrought iron can be said to represent an Afro-American tradition in blacksmithing. Euro-American ironworkers might have similar experiences but that does not weaken Philip's ties to the deep cognitive patterns that are basic to black culture. The improvisatory design process which he employs is not required by his technology, forms, or materials. We can account for it only through his unique personality or his cultural orientation. The similarities demonstrated between Philip's work and that of other black folk painters, musicians, sculptors, storytellers, and other artists suggest that his ethnic heritage cannot be disregarded as a significant element in his ironwork.

5

"I Done Try 'Em All"

PHILIP SIMMONS is more than a blacksmith, even though that is the label he gives himself. While the public generally recognizes him for his craft skills and artistic talents, his life is not just a record of ironwork. He is also a family patriarch, a churchman, a Boy Scout leader, a sportsman. To overlook these elements of his biography would be to paint a portrait without shadows and thus give a rendering that lacked depth and sensitivity. Philip the blacksmith is an active member of his community and he serves that community in many ways.

Philip's parents died while he was still a young man. His father William died shortly after Philip was apprenticed as a blacksmith. As the oldest child, he then became a male authority figure for his brother and sisters. They called him Big Bubba and thought that he could do anything. They occasionally tagged along after him down to the blacksmith shop and watched in awe from the door as he performed his appointed duties. With pride they pointed out to passersby that their brother was the " 'prentice." Philip recalls that his sister Mary often solicited his help with her school work and thought so much of his seemingly bottomless store of knowledge that she once came to him to have a sore tooth pulled.

As Philip matured in his trade, he worked longer hours, often past dinner. His brother Edgar, nicknamed "Skeet" (short for mosquito), remembers carrying Philip's supper down to him at the shop. Peter had usually gone home, leaving Philip behind to do things on his own. Skeet would deliver his parcel and then lend a hand if he could, or just sit and watch. He did not carry on in the trade but claims to have gotten some experience in blacksmithing. No doubt Philip passed on to Skeet some of the knowledge he had acquired from Peter. Some of this after-hours smithery was surely a big brother showing off, but some of it was just as surely a young man assuming a paternal role. That role was finally confirmed at the age of twenty, when his mother died.

Four years later Philip married Erthly Porchie, who was then only seventeen, and in just four years they had three children. At the point when his life seemed so full of promise, with a good business and a growing family, Philip was struck with an unexpected tragedy. His wife became so ill that she needed to be hospitalized. Philip visited her each morning for six weeks, and Erthly's condition improved so much that the nurse said that "she'd be out in a week." Philip was at his anvil later that day when an undertaker came by to give him the word that Erthly had died. Suddenly his family was again shaken. Philip wondered how he was going to raise three small babies and run his shop. He decided that he needed to work so that he could feed, clothe, and educate them. They were each placed with a relative or a friend.

Mrs. Brevard took care of Lillian. One sister took the boy and the other sister took the other girl and care for them. That's give me a chance to get out and work for them.

The extended family network of kin and neighbors absorbed his offspring when the nuclear unit was broken. The children were not only cared for and given the affection of a close family circle, but Philip himself found a way to perform his familial duties. In a way, Philip's family grew. He was now a lone father with doubled kin ties to his sisters' families plus a strong friendship bond to a kindly neighbor. His paternal image thus expanded over three different locations, whereas previously it was concentrated in one place. This situation pleased the children and suited him.

They had a mother . . . needed a mother, so maybe that's why I didn't done marry maybe. 'Cause the children was very comfortable, maybe that's why I ain't married since.

The leadership role he had played as the big brother was in a sense recreated by the death of his wife. Once the children were cared for, he redoubled his efforts to do well as a blacksmith. Then on his occasional visits to each household, he appeared both as the fatherly provider and the successful craftsman. His status was distinct, for he belonged to no specific family, but to several families. In fact, from 1940 to 1958 he changed his personal residence six times. He was inside the family and outside it at the same time; he was in a sense above the usual kinship rules and thus was respected by many rather than just a few. The first community in which his reputation was to grow was then the community of the extended family.

Over the years Philip's industrious behavior has embellished his reputation among his kinsmen. His apparent economic success makes him seem, in their eyes, a notable example worth following. Nephews and young cousins are encouraged to imitate Philip's resourcefulness, to get a trade and to work hard at it. Thus, family respect carries with it the demands of family responsibility. Because Philip seems capable of helping others, his aid is readily sought. Family members may come for advice on a bank balance, for a loan to pay overdue bills, for a job in the blacksmith shop, for an opinion on matters of Scripture, for a recommendation to a prospective employer. Philip is now acknowledged as a patriarch and as an arbitrator on all sorts of family matters. He is still the same leader he was as Big Bubba. He grudgingly admits that his willingness to solve family problems can at times be a bother, particularly when others fail to consider simple solutions for themselves: "Some people wouldn't buy a box of matches if they thought it was too complicated, rather I would do it for them."

Mundane matters intrude into the work of the blacksmith shop. Philip is more and more often called away from the forge by demands which he cannot refuse because of the bonds of kinship. Given the nature of blacksmithing, which requires a steady, even work pace to complete a piece of wrought iron successfully, it is important to realize that Philip's career has developed as his role within the Simmons clan and the East Side neighborhood has become more and more complex. He has had to negotiate social crises simultaneously with the intricate problems of artistic design.

As far back as Philip can remember, his family has belonged to the Reformed Episcopal

Church, and he has followed that custom. He has been an extremely active member of St. John's congregation on Anson Street. Shortly after his wife's death his participation increased markedly. He took part in programs involving the training of young boys, namely the Y M C A and the Boy Scouts. For over thirty-five years he has led the church's Scout troop and gained a number of citations and commendations for his service. One of the activities he stages for them yearly is a visit to his shop, where he demonstrates the techniques of blacksmithing. The object of the visit is "to show how important it is to have a trade. Whatsoever it is, isn't so particular. But if you learn about iron you can go so many places and that's is important for a young person to learn."

When he teaches Sunday School he again makes use of his profession, at least in his own mind. As he prepares lessons from a standard textbook he thinks: "Raw iron ain't worth nothin', you got to shape it for it to be something. Same with a child, you got to make something of him." But his religion also extends into blacksmithing. Once when he was asked about the origins of the craft, Philip asked the questioner to remember the Bible, particularly the Crucifixion of Christ.

There always been a blacksmith from way back, you know since Christ's time. I was told that a blacksmith made it [the nails]. So then you can go back as far as you want . . . so the nail, they had the original nail. I cannot tell you beyond that.

As a churchman Philip has held several positions: member and past president of the Usher's Board, member of the Vestry Board, and treasurer of the Savings Club. Currently he is treasurer for all the churches in the local synod of the Reformed Episcopal Church, as well as treasurer for St. John's youth group. The responsibility for financial matters has probably been granted to him because of his proven ability as a local businessman. In the shop he must deal with a host of fiscal concerns ranging from taxes to insurance to license fees. The monetary matters of the church amount to more of the same kinds of bookkeeping tasks. Philip acknowledges that it all requires extra hours of his attention, but since he knows how to handle such matters it is fitting that he should be charged with keeping the financial records.

In addition to the formal duties of church administration, Philip also devotes some time to unsolicited acts of kindness which he refers to as "charitable work."

I visit the senior citizen. I tape-record and I record the service every Sunday and I take it to the shut-in. . . . I get a kick out of that up until today. I still enjoy it. We went to General Council [in Chicago]. I tape the whole Council, you know. And one man was going to General Council all his life and he's about seventy years old now and he'd give anything to hear the General Council, the proceedings, so I record it, I'm going to take it to him. That's going to be a surprise for him . . . he was going General Council about thirty years and this last one he was shut in, sick and shut in, and couldn't go. And he didn't know that I record it special for him. So I goin' enjoy that.

On another occasion he modified a parishoner's mailbox. In this case a woman was bedridden and worried that her pension check might get stolen before she could retrieve it from the mailbox. To set her mind at ease, Philip devised a lockable flap that fit over the slot in the box. The task occupied about two hours in the middle of a working morning, but he didn't charge for his time or materials. He only commented when he had finished, "Now maybe she'll rest easier." He estimates that about once a week he does a job for no charge. Those activities are, according to Philip, the requirements of a good Christian life.

But a moral life is not without its lighter, relaxed moments. Philip's main recreation is sports. He once advised his nephew Carlton, who is currently apprenticed as a blacksmith, "You've gotta give up your playin' time. I did, no checkers." But he added later, "except baseball, wouldn't work when there was a baseball game." Philip was an avid ball player. As a young man he played on a regular hardball team. The games raised such an interest in the community that even the stern Peter Simmons would close down the blacksmith shop on game days and go and watch his apprentice play. This was probably made easier by the fact that for a while many games were played in an empty lot near the blacksmith shop. There was a series of teams that played and, while they were certainly amateurs, the caliber of play was good enough to attract regular visits by professional teams. Philip recalls those glorious days with obvious delight.

I played baseball. It was organized 'cause we had uniforms and we had a team. We had a league right here in Charleston. You could call it professional in one sense, now we call it

Pony League. When you get organized then you join the league. Then you can play in the park. People pay admission to come in. That particular time we didn't have A league, A and B league then. We only had the professional. That time when the Cuban All-Star and Black Yankees and the Kansas City Monarch and all the teams. They came down here to play.

I played left field. I catched till I got my arm hurt, loosed my throwing ability so then I played left field. Right [field], center [field] sometime. I played for the Diamond Sluggers. We have six teams around here. East Bay Sluggers, the Diamond Sluggers, the Red Caps, Charleston Eagles, the Mexico Wild Cat. We had several teams. We mainly played each other. We didn't travel. The team that traveled was the team before us, the Charleston Piedmont. They were in the Negro League. We were just local. We played against them, our team was good. When they [the Charleston Piedmont] home they don't have nobody to play, they play against us. If it was now time, they would be playing for the big leagues. You know the Kansas City Monarchs and the New York Black Banner and them team, on exhibition they used to play the Yankees and beat them too. At that time the teams didn't have no blacks on the team.

We play those professional teams. Right now I can think of the Savannah All-Star. They used to play from, start down south and play on up the early part of the season in April, go on up. When they'd come through we'd play against them in a large park too. They'd go on up and play the Yankees. Start from down with the Jacksonville Eagles and all them. They was good teams and all, but they just couldn't go in the majors. We talkin' back there in the thirties then. We were grown mens. Then we were in our twenties then. I played more baseball after the death of my wife. When I got married we were playing ball then because see I was [at that] age then. I started playing baseball at the age . . . about fifteen. Then I start playing with Diamond's club, think I been about between seventeen and nineteen years old. And that was the best part of my life, between twenty-one and thirty.

The only chance we got to play the pros is when they come up here. They'd be one team after another. Sometime it'd be for months. We'd have a team there every Sunday, a new team every Sunday.

Philip played on the hardball teams until he was thirty-five years old and then tapered off to more informal games of softball at the church's summer camp or outings with the Boy Scouts. Even at the age of sixty-seven he continues to participate in the sport as an umpire. He claims to have "shortness of breath for running." Otherwise he would still play left field as he did for the Diamond Sluggers.

As if the work of a blacksmith shop coupled with the responsibilities of a family, a church congregation, a neighborhood, and a baseball team did not occupy his every waking moment, Philip also experimented with other business enterprises. One of these was taxi driving.

You know it wasn't a permanent thing. You know the driver when he get tired instead of hiring another driver, I do a little driving. I owned the cab, drive at night. But I didn't drive it too often now. . . . I bought the cab, my friend want to drive it. He had a cab, he was renting a cab and a guy took the cab. So

I said, "I'll swap my car 'cause I don't need the car. If I want to go any place, I'll call you." So I swap my car and get cab for it. Might have even paid off, but I took it off the line. That's been in about '48. I was in there about a year and six months, that all. Too long. Didn't take me anytime but the guy was driving 'em, he gamble out the money. He fool out the money. So I just get disgusted 'cause I was helpin' him, so I took the cab over. But he drive all day and at night he say he want to go home. I take the cab and drive about three, four hours. Grab a few pennies.

Philip then experimented with two other money-making schemes.

After I got out of the cab business, I got a cleaning business, me and a friend of mine. Both of us put up the money. We didn't have too much money either. That's how come we went bankrupt. We went bankrupt on this one, not the cab.

First it was the restaurant business. I had a restaurant down on the east side of Marsh Street. All the equipments, right down there to the foot of Calhoun. . . . Last about eleven month I remember. Our cook put us out of business. He said, "Well, I know you all ain't makin' no money. I'll cook for you." He cook. Well, business wasn't paying off so he couldn't get a raise. So he quit. When he quit then I say, "Well, George, you'll cook, cook all our food tonight." [He] said, "Well, that's it, you got to hire [someone] to help me clean up." He was doin' the cookin' then. You run the restaurant with a little help, a little unexperience' girl. So he had all the experience. George'll go to this other guy. So you know what happened? We couldn't compete with the other fellow. We had to go out of business.

I never got out of the blacksmith meantime. But I wanted to help myself. You know the blacksmith bring [just] so much and I say I could make some extra money on the side. I done try 'em all.

These failures made one thing extremely clear to Philip: his future lay in blacksmithing. Ironwork provided a steady, if modest, income. Speculation in side lines only wasted or "fooled" the money away. With a family to support there was little margin for such gambles. His dalliance with cab driving, drycleaning, and restauranteering directed Philip more strongly to the forge and anvil. With these adventures of the late 1940s behind him, he took full command of Charleston blacksmithing. By 1950 he was the only blacksmith listed in the Charleston *City Directory*.

While Philip spread his energies throughout Charleston in several capacities, his shop remained the major point of his interaction with the community. Customers came in for tools, repairs and modifications of cars, and for decorative ironwork. Young men came in for jobs. In the past forty years many potential apprentices have come and gone. Most liked the glamor of the flame and forge, but soon gave up when confronted with the drudgery of hours of preparation of material and the final grinding and wirebrushing of gates and fences. Philip often depended on "picking up" someone to help him with a piece which was too big to lift by himself. He noted that there "was always someone on the street who needs a job. I just pay them for a few days, then then go down to the wharf, stevedoring. A few weeks

later maybe they come by again." Often the part-time "helps" were members of Philip's extended family: cousins and nephews. Today his brother, nephew, and grandson will occasionally pitch in when their aid is needed.

I had 'prentice off and on. I always had somebody clean up the shop. And after they grow up they become a permanent fixture. Just like Si [Sessions] and Ronnie [Pringle] and a cousin of mine, Dan Alston, Herbert Alston, Joseph Green, Simon Alston. I had a lot of my family, too, just like Pringle. Ronnie is my first cousin. Si and them other boys they no kin to me at all. They were in Scouting and they come there to earn a merit badge.

The Boy Scouts, kinship ties, and a community reputation provided Philip with the supplementary labor required to run his shop. These broad-ranging social relationships helped support the craft tradition.

Ronnie Pringle, in recounting his entrance into blacksmithing, confirms this pattern.

The first time I started working in the shop, he was scoutmaster. I was in his troop. We were supposed to have a meeting the next day but we didn't know the time. He was doing something then, I think a fence, and he needed a hand. So he asked us if we would mind doing it. We said, "Sure." We didn't have nothing else to do. So we turned the gate that day. Come back the next day. Then again, then again. Started hanging around when we didn't have nothing to do. Looking at him work on the forge on Friday, fascinating work on Friday. So we started getting interested, taking a raw piece of

material, turning it into something beautiful
you know. . . . When you finish, the
satisfaction you have—I just can't explain it, I
just feel it. I been with him ever since up till
now.

Ronnie and his close friend Silas Sessions
both started their visits to the blacksmith
shop at the age of twelve, in 1955. As eager
youngsters they must have reminded Philip of
his first fascination with Peter Simmons's
exciting work. They have for the last two
decades been the most steady helpers. Except
for three-year hitches in the military service,
they have always been in the shop and have
now assumed a full partnership in the trade.
Philip considers the three of them to be a
team. When he was invited to attend the
Festival of American Folklife in Washington,
D.C., in 1976, he only went when he was
assured that his team could come along too. Si
and Ronnie are two of the primary heirs of
Philip's craft legacy. Others have received the
training of a blacksmith only to go into other
trades. Si and Ronnie are in fact no longer
apprentices, but full-fledged smiths. Yet
recognizing that Philip, the "Boss," has
superior knowledge of the tradition, they still
consider him their leader.

The apprentices did not always turn out as
well trained as Si and Ronnie. Some boys,
through their enthusiasm and imperfect
control of the technology, created problems for
their master. Philip recalled one anecdote to
illustrate some of the difficulties that arose
periodically. A customer had left his car to
have a running board attached.

The boys didn't done a good job. They rush it
before I get back to the shop. Finish it before

I get back to the shop. The guy went up the
street. Pick up his wife and he wife got, open
the door, got on the running board, the whole
thing fell down on the ground. So he brought
it right back. That's the same day too. No, the
next day . . . came back early the next day.
When he came in the shop, he say, "Where
your boys?" I say, "They isn't here." "[They]
fix this running board yesterday. Wife step on
it. It drop down." I say, "I guess you want 'em
to go back over it." "Yeah, I likes for 'em to fix
it back." "Okay."
He wasn't mad. Wasn't mad at all. So I got
on it and fix it. Have to fix it for free. I weren't
mad. Didn't feel good with the boys 'cause
they hadn't done a good job. When the whole
thing boil down, you know what happen?
They done it in the shop and I was
responsible for anything that go out of the
shop. And the easiest way for me, is to get on
the running board and fix it. Now if I was
there. . . . They charge the man four dollars. If
I was there I would have charge him twelve
and done a decent job. So I had to get on it
and I done a twelve dollars job for free. If I
had fix it for the price of four dollars, I would
have done the same thing. So I would fix it as
though it was me the first time.
So you see it cost [me] but they didn't give
it a thought. They just think that was fun for
them. Turn out a job and make their own
[money]. And they made their own price.
Those boys was unexperienced. I think they
had been coming in the shop two years and
doing little things on they own, practicing how
to do things, so they feel themselves, think
they know a little somethin' 'bout the work.
Then they launch out to take these jobs. Take
these responsibility on they self. But most
time they can't quite.

74   Philip Simmons and Carlton Simmons.

75   Ira De Koven.

76   Philip Simmons's present shop, 30½ Blake Street.

Such were the master craftsman's trials as he sought to bring his apprentices to their full development. Philip reminds his current helpers that what they have to learn now is easier than when he was an apprentice.

We didn't have no 'cetylene [torch] to cut [with]. So you see, we got all these modern equipment. That's where the helper comes in there. The 'prentice comes in the shop now on the easy end. [In the old days] we have to heat it [iron], turn it [the forge] with our hand to make that fire. The Old Man'd take the chisel and put it on the iron, [you] come an' take the sledge, you finish, go back there and turn it. It's hard then, real hard.

The training he offers includes the older, more difficult means as well as the modern techniques involving electric arc welding and acetylene braising. In this way if a smith should run short of his supplies he could still function. The oldest practices, according to Philip, are "a hard way to do, but a nice way to know."

By teaching the processes of forge welding, upsetting, punching, riveting, and other handwork aspects of blacksmithing to his apprentices, Philip recreates the phases of his own training. Indeed, he brings the nineteenth century further into the twentieth. Moreover, he reaffirms his ties to the community by giving it not only the benefit of his own skills, but also more practitioners of one of Charleston's most venerable crafts.

In the last three years two more apprentices have taken up the trade: Philip's nephew Carlton Simmons (plate 74) and Willie Williams. Another blacksmith, Ira De Koven, a university-trained ironworker, has also regularly solicited Philip's advice on several commissions. When Ira made his first balcony, Philip in fact helped him anchor it to the wall (plate 75).

Philip Simmons is a teacher. His life of good deeds provides an example for many to follow. His lessons in Sunday School lead some into the mysteries of religion. His Scouts acquire the skills of self-reliance and the moral sense of duty. His apprentices learn to be blacksmiths. By giving what he knows and what he does to them, Philip ties himself to a social network. His artistry has not kept him aloof, but provides a means of serving his family, his neighborhood, and his city. He once expressed his relationship to his community in this manner:

I owe all my career to the people of Charleston. Without they giving me the chance, I couldn't have anything. I can't make a gate if they don't want 'em.

6

"I Could Do This Work Just Almost for Free"

PHILIP'S INFLUENCES have clearly molded the lives of his apprentices, at least those who have stayed with him. In their evaluations of decorative work they employ many of the same terms he would use. Si Sessions (plate 77) emphasizes the need for smoothness of finish.

You take the rough spot off, clean it. Ain't nothing look worse than making something beautiful and have it all roughed up.

He also adds that designs must have a degree of openness:

When you're working with iron, it's not too good to have the design too clustered together. That take away from the whole work altogether. A bunch of one thing in one of those spot, one of those spaces, like trying to put five thousand match stick in a little box.

Ronnie Pringle (plate 78) tells the same story as Philip concerning the shape of leaf decorations.

Every time we make one, we make different design. It's always different, very seldom look the same. No two leaves on a tree'll be alike, so no two pieces of that'll be alike.

Both men have learned to develop designs by constant sketching. Si's experience again closely follows Philip's.

Normally customer want a gate—we draws three or four, five different kinds of gates and let him look through and pick the kind he wants. Whatever one he choose, that's the one we have to make him. You know when I be home, I go to bed at nine o'clock. I probably wake up about twelve, cannot go back to sleep so I get some paper and pencil and I started drawing and pretty soon I go to sleep. I may not finish my drawing that night, but maybe the next night I'll finish the drawing.

After more than twenty years of training and partnership with Philip, Si and Ronnie find it difficult to recount the events of their apprenticeship. In some ways it is still going on, although now, in their mid-thirties, they are proven professionals. Their training seems like something they have gotten beyond and they rarely think about their days as novices. Consequently neither man had much to say about apprenticeship. Fortunately Willie Williams (plate 79) is in the midst of his training and can tell us just what the process is like. He can say why he wants to become a blacksmith, whereas for Ronnie and Si the choice has long been made.

Willie was born in Charleston in 1950, and, ironically, for the first eleven years of his life he lived in the housing project very near Philip's Laurens Street shop. Willie recalls seeing the billows of black smoke from the forge drift toward the sky, but he did not venture over to investigate further. Willie left for Philadelphia at the age of twelve. His

78   *Ronnie Pringle.*

77   *Silas Sessions.*

79   *Willie Williams.*

main memories of life in "Philly" revolve around school work, especially art classes. He claims to have enjoyed drawing and consequently did poorly in other subjects because he "favored art." Once he won a prize for some art lessons, but his father wouldn't allow him to accept them. Later his artistic muse found expression in sculpture. Willie found a burnt piece of wood in the furnace that already seemed to him to have the suggestion of a face in it. Using a screwdriver and a brick, he transformed the charred board into a small head. After high school he turned his attention to more practical interests and took a job in the Sun Shipyard in Chester, Pennsylvania. Thus, when he came eventually to try blacksmithing he already knew something about working with metal, besides having an artistic disposition.

His apprenticeship began with his return to Charleston, and his reasons for coming home run deep.

Suddenly decided to come back again and start my life over again in Charleston. I noticed that the society there didn't have basic things of life, important things such as hard labor. I mean basic living: fresh air, sunshine, trees. Of course they're up there but not in the neighborhood where I come from. I came from the ghettos of Pennsylvania, east Philadelphia, and the only thing I knew up there was street corners.

So one day I just decided to come back home. I told my mother I was fed up with city life. I came into somewhat of an awareness of how very important it was to be around fresh air and trees and just old country living. However, I'm not calling Charleston country

living, that the people are countrified. But I think that the air is a lot more pure here than it is in the ghettos of Philadelphia and New York which I was raised in. It's made a complete change in my life. I enjoy just basic things, socializing with people. No more hustle and bustle and people stepping on your toes in the bus and stuff like that and subways and that kind of stuff. It's different.

Evidently Willie was ready for a major change when he arrived in Charleston. His dislike of Philadelphia's environment is probably a metaphor for a spectrum of social circumstances involving employment, family, and peer group. All of these matters have now changed for the better in Willie's view, since he regards living in Charleston to be a "triple bonus."

I think I've been blessed to come here at this time and run into such a reward here. I think it's a blessing just to be here at this time, being able to be taught by this man, Mr. Simmons. I think it's a blessing just for me to be here.

Certainly some degree of good luck has played a role in Willie's arrival in Philip's shop. Willie had purchased a welding machine and a truck before leaving Philadelphia and was taking on random repair jobs around Charleston. One afternoon, while considering who might provide him with his next job, he was approached by a local minister.

[The minister said] "I know a man that you might be interested in." I said, "Well who is it?" He says, "Philip Simmons." I said, "Well

I'd love to meet him, I've heard so much about him." Came here, the reverend brought me here . . . and the minute that I saw what they were doing here I decided to stay. I been here thirteen months and I been coming here just about every day, practicing on the anvil. Always loved working with metal, but I never could seem to find the right teacher or right thing that I felt suitable in. It's okay to be able to weld metal together, but it's another thing to be able to bend it, to shape it the way you desire. When I came around here and I saw some of the stuff that Philip Simmons was doing . . . and I asked him in such a way to accept me into his shop that he couldn't refuse. It was like a child asking a father can he go out and play. I wanted it so bad it was just like me looking up at God or somebody and praying. This is the way that I took it.

He saw that I was sincere and this is what he said: "You seem to be a sincere person. This is why I really be glad to have you in the shop." And I'm sure that I was sincere because for thirteen months I've been practicing every day and I can see the rewards of it.

One of the prime rewards for Willie is the opportunity to create his own works of art—something he had not had a chance to do since his school days.

When I first came into the trade or the art I was willing to do it for free. You know, come around after work and do it for free. This is how bad that I wanted the art because I noticed this was something entirely different from the line of work that I been doin'. It was a chance to express my inner self. Like I said,

it's good to know how to read a ruler or do mathematics or weld steel together or burn steel, but it's another thing to design—stuff like that, create your own things. This is what I enjoy. Just being able to come from my mind and put it into reality. This was the chance in a lifetime and each [time] that I'm around I feel that I'm coming closer and closer to harvest.

One of the main things is you take iron or metal. It's the hardest thing in the universe. But once you heat it, it becomes one of the softest things in the universe. Then you can shape it. It's just like being the god over a piece of iron or something. Even though I know I'm not God, it's just being able to take something from nothing and turn it into something beautiful. Create something from nothing which I think is remarkable. Pick up an old piece of metal, old rusty metal just laying around. And throw it in the heat, bang it a few times, and here you've got a chisel. Something that you can go out and make money with. This is what I think is remarkable about the trade: making things or making the iron bend to your will.

I consider this my only profession even though I have a part-time job at the hospital. One day, hopefully, I can give all other work up and just apply myself to just this one thing. I feel somewhat of an artist. I feel I belong more to a blacksmith. Just a thing that I want to do.

Willie's commitment to the trade was shown on numerous occasions when after working the night shift at the hospital, he would report to the shop to put in another eight to ten hours. His energy was amazing. Indeed, he often seemed like a man driven by deep emotion. He also has a commitment to Philip and is grateful that he was taken into the shop.

I'm still an apprentice. We work on commission. I have my own tools and we work on commission basis. However, at times I feel my take is a little bit too much. But Mr. Simmons is a type of person, I call him a fair man. A lot of times he's paid me for jobs and he's asked me, he says, "Do you feel that that's enough?" And I say, "Well, I feel that I owe you."

Willie's firm resolution to make himself into a blacksmith has made him an eager student. Philip has even commented that he has never had an apprentice learn so fast. (He also never had an apprentice who was already a mature adult.) Willie has compressed six years of learning into a single year.

When Willie came for his first day's work as a blacksmith, Philip decided to test his resolve with an unappealing task. He directed Willie to pick up a sledge hammer and straighten a driveway gate that had been damaged by a truck. Usually a bent section of iron is heated until red, after which it can be more easily brought back to its original shape. Since the gate was to be repaired cold, the metal was unyielding. It took Willie most of the day to finish this job, and when it was completed Philip knew that the drudgery of blacksmithing would not deter his new helper. Willie's version of his first day is, as we might expect, slightly different.

One of the first things I did in the shop was to learn how to straighten out old beat-up pieces of steel. As I can remember, one time Mr. Simmons had got a gate to repair that was completely crushed. I mean smashed from one end to the other. And he said, "Now's a good chance for you to learn how to straighten steel out." And I couldn't see the worth. So I fought with this gate, and fought with it, and fought with it, and sweat and sweat till finally I straightened that gate out. And now I can see where it was absolutely mandatory that I learn how to straighten steel out because every piece of new steel that you buy is bent. And you have to know how to look down at your steel and see whether it's straight or crooked. That's one of the first things you do when you put things together in blacksmithing. So even though I was too young in the mind or too young in the area of blacksmithing to see the value of that, Mr. Simmons saw it all the time and this was, I guess he was just giving me some experience even though I didn't like it. That gate almost caused me to quit.

From that point Willie began to acquire the other skills of a blacksmith.

Mr. Simmons believes in breaking things down to you to its lowest term. I think because of the fact that he's such a master he knows just what level to start you on. Say for instance learning how to straighten out steel from the beginning. Then learning how to start a fire, to keep it burning. These are some of the beginning things of our art or trade. These were things that were very hard for me to learn, but they were absolutely mandatory for me to know 'cause if you can't keep a fire burning then you can't keep your iron hot. You can't get it hot enough to bend. If you can't straighten out steel from the beginning, then all your work comes out crooked. You have to have a good eye when you look down your steel to see whether it's straight and then you have to have a nice fire.

So he starts you at the level where he feels you fit in. He's able to. He knows just where to place you at the right time. A lot of times I thought it was too much for me to handle, you know, at that particular time. But it was just that I had an inward doubt, being in this new field and not knowing the names of the tools and not knowing how to hold the tools and scared of the fire and everything. And getting your hands burned and everything. I mean I went through all the basics and suffered all the scars and sometime I still hurt myself 'cause I consider myself still a beginner in this field. But if I must say so myself, I've advanced quite a bit. The simple reason for that is that I had a good teacher or that I have a good teacher.

As part of his training Willie has acquired not only basic techniques but an added appreciation of old-fashioned ironworking techniques.

Well the main goal I would like to reach in this art . . . I've seen the old works here in Charleston. I mean the old rivets, the way they were put together hundreds of years ago I say. The old-fashioned way, the way that they used to do it without modern-day equipment. As a matter of fact, every time I get a chance to buy a piece of old blacksmith tool or any other, I buy it and hold on to it. No matter what condition it's in, I hold it because I

80   *Philip Simmons and Willie Williams.*

would like for people to have the experience of coming and seeing an old world or a way that things used to be done many, many years ago. That's a beautiful thing to see. And I would like to master the art the way that the old masters had it many years ago. It's like reaching back for the past and trying to grab it.

Well it's just . . . well, take riveting. A person of that day many years ago, the old iron masters, they may have five hundred pieces to rivet whereas today I would just take a welding machine and weld the five hundred pieces together. They would take their time and drill a small hole in each one. Five hundred pieces and run a rivet in and seal it up. And I can see where it took a lot of time and a lot of patience to do this, which is not done today. Everything is just quick, quick like that. This why I say that I could do this work just almost for free.

And one day this is how I'd like to do it. I'd like to do it just the way they did it years ago. I don't intended to try to get rich off this trade. I would just like to have the trade to be able to pass on to the younger ones and just keep it alive, keep it from dying.

Willie, like Philip during his youth, has been positively influenced by the Charleston environment. The numerous examples of decorative ironwork provide him with models and inspiration. The nostalgic "feel" of the city has rubbed off on him. As he travels through the city streets he keeps an eye peeled for noteworthy fences and gates.

There are quite a few [gates] and many I don't know they names of. But I'm gonna tell you, well, Mr. Simmons have quite a few of 'em. Mr. Simmons has one in particular that I really love, which is the Snake Gate. There are many [gates] here. Some are made by Germans, some are made by blacks. But there are many of 'em. Well I turn my eyes backwards in my head as I ride by just to take a look at it. There's quite some beautiful work here in Charleston.

After six months of learning the basic skills of blacksmithing, in addition to developing a critical eye for the local ironwork, Willie took his first commission.

The first thing I got a commission on was window guards, window grills, which is a fairly simple thing. That's one of the first jobs I've got here in the shop. And then we went on to [hand] rails. Then we went on to large gates—well, semilarge gates. I exaggerate some.

After only a year Willie progressed so far that he could contract for his own jobs. This was a remarkable achievement. In a relatively short time the faltering novice had become a capable journeyman. He was now doing his own designs. While he still cleared his procedures with Philip, his actions were pretty much his own. Philip took on the role of foreman and checked Willie's work from time to time in order to give helpful directions. When Willie was assembling his first driveway gate, Philip reminded him about the principles of design that he himself favored. He warned Willie not to make his scrolls too clustered and to ensure that they were properly rounded, he told him to "make a circle." Later, when Willie fashioned an open spear point out of a slat which was to go on top of the gate, Philip gave him some necessary criticism: it was "puffier" on one side and had "too much belly." Philip told Willie to return to the forge to reheat the piece and reshape it. In similar fashion Simmons advised Willie to rework some of the scrolls because they were "too small and squatty." When Willie placed them into the gate they were to be set on a descending line running from the center to the outer edge. Philip noted that Willie had not allowed for enough slope. Without telling him exactly where the scrolls should be placed, he suggested, "Drop down like you coming down a hill." With these comments from the master, Willie completed a gate of which he was proud, but noted that it was a good piece because he had done it "in the Phil Simmons style."

Philip says that it is best to teach by example. "That way," he adds, "your brain don't get cluttered by directions, brain block up." He prefers to demonstrate how a job should be done and thus give a clear image of process, technique, and finished product. When Willie finished his gate, Philip accompanied him to the site of installation and took charge of setting it up. He showed Willie how to set the gate anchors in brick pillars on each side of the gate, how to check the gate for proper swing, and how to avoid the unsightly sag often found in a double driveway gate.

On another occasion when Philip ran short of acetylene for his cutting torch he decided to take the opportunity to demonstrate how to punch the required hole by hand. First he upset the bar, giving the butt end greater thickness and consequently ensuring that the hole punched there wouldn't seriously weaken the bar. Next he placed a handful of coal dust on the beak of the anvil. When the upset end of the bar was heated red, he placed some of the dust on the bar and struck that spot with a hammer and punch. The coal dust, he explained, keeps the punch from adhering to the hot iron. After punching halfway through the metal he then plunged the punch into a bucket of water to cool it down so that it would stay hard. He also noted that the end of the punch was becoming dull, and he quickly tapped it back into shape. Then he turned the bar over and punched it through from that side. Soon a small, flat disk dropped out of the bar and the hole was formed.

He repeated these actions a second time before turning the job over to Willie. When Willie began, Philip urged, "Go ahead, don't

81   *Philip Simmons showing Willie Williams a forging technique.*

stop, go ahead." Once the hole was started and Willie was checking the condition of his punch, Philip ordered: "Cool it off. Hurry up, don't take too long about it." But Willie was too slow and the bar cooled below punching heat. While the bar was reheating on the forge Willie was directed to "take your punch and taper it . . . not too hard. Just only knock the burr you put on it." When the bar was ready again Willie noted, "That's good," and began to hit the punch, but Philip observed, "Seem like you ain't got enough grip there or something."

"Enough grip?"

"I mean with your hand. You ain't puttin' enough weight." After a few more blows Philip continued: "Too cold, that's too cold. You should put that through with one heat."

"Think so?"

"Definitely so. Put it back in the fire."

While waiting for the bar to get back up to temperature Willie proceeded to dress the punch as he had done earlier. Philip now approved: "That's right, knock it just on the end." He then added: "Put it in the water and cool it off. Now when you put that punch back on the rack, the other person come in behind you, he got a punch to work with."

At that point Willie turned to the forge to retrieve the bar. Philip cautioned, "Wait a while," looked at what his apprentice was about to do and continued: "That's all right, you got it on top of the forge. Don't put it down in the bottom. You don't get nothin' but air in the bottom. Put it on the top." When the bar was again a glowing red color, Willie placed it on the anvil and began to punch it through from the opposite side. All the while Philip, standing beside him, watched his work,

and when the two halves of the hole didn't align properly he added: "You got to come straight down with your punch. You got to come back in the other side, straighten that out." He advised Willie to drive the punch through while the bar was still hot: "A little more on the other side, a little more. Hit hard, it won't hurt it."

Obviously, example alone was not enough to convey the many steps involved in the task of punching a hole in a square bar. But in this lesson Willie had not only learned how to make a hole, but how to care for tools, how to add thickness to an iron bar, how to grip and swing the hammer, something about the ability of certain sizes of stock to hold the heat, and about the need to act quickly. This was a lot of information to digest. Philip noted, "Maybe he will pick up more just in this hour we were here than he will pick up in the other 365 days."

After spending most of 1977 as an apprentice, Willie thought that he would still need two more years of training.

The way that I figure, it should take about two more years until I can learn all the basics here in the shop. I mean really familiarize myself with all the tools and all the use of tools, learn how to forge weld, and learn how to make many more shapes. Learn how to shape different things. At the end of that two years I should be pretty well implanted here and then I should be able to branch out and do what I really want to do.

But Willie's progress was faster than he expected. Within six months he had started his own shop. He hasn't severed all his ties with Philip, though. He visits and consults often on the fine points of blacksmithing. Philip, in turn, sends over to Willie work that his shop cannot handle because of his backlog of work requests. There is thus an interaction between the two shops based on the initial master-apprentice relationship. Willie feels indebted to Philip and, more important, senses that there is still more to learn. In the meanwhile he has joined the very special corps of Charleston blacksmiths. Along with Silas Sessions, Ronnie Pringle, and Ira De Koven, he will carry the trade well into the twenty-first century. At the same time he and his fellow workers will perpetuate the example and influence of Philip Simmons. They have all been influenced by "Phil Simmons's style."

7

"Look Backward to See Forward"

THE DEMAND for a blacksmith's skill remains strong in Charleston today. One might even surmise that the opportunities for decorative work have increased, since many of the requests are for interior ornament not commonly used until the last few decades. Thinking back on old Peter's advice not to worry about the future of the ironworking trade, Philip can now see better than ever that the Old Man was right.

I don't know what he was talkin' about 'til now. Look at the work we done inside the house the other day. The interior, inside, not the outside. You can see the work on the outside. She [the client] done her whole stairs out of wrought iron. Take out the wood, put wrought iron. Like the man over here, he name is Green isn't it? He made a wrought-iron mantel and he sent off and get a marble—I don't know, a heavy marble [slab]. That thing must have cost about maybe three or four hundred dollars. Maybe more than that—one inch thick, twelve inch wide. [I] made a bracket and put on the wall and set that mantel on it. And he took his wooden step [rail] out. It a new house he had 'cause an old building wouldn't have a wooden step [rail]. Took that, it all out and

come up here on the side, turn around and go up the corner [with an iron railing]. So you can see it a big demand now just to build a house and complete 'em inside. That's plenty of wrought iron, you know.

The present is thus, in Philip's view, the best of all possible times for blacksmithing, at least in the decorative field of the trade. While it may strike us as strange that a nineteenth-century craft would be able to flourish in the late twentieth century, we can certainly understand Philip's optimism. Presently he has command over most of the hand-wrought decorative ironwork in Charleston, and the demand for it is constant. He has to worry not about finding work, but rather about how to schedule all the jobs so that no business is lost. For a man of action like Philip such circumstances fulfill the necessary conditions for a busy, prosperous, and meaningful life.

Optimistic and promising business conditions promote a general spirit of hopefulness in Philip. Even after more than half a century of blacksmithing he still, on occasion, will see himself as a mere beginner, an apprentice fresh on the job.

And you know about this thing, I'm learning too now. After those many years, after fifty years I'm still learning things. I'm doing things and I have done things and I'm still doing things I have never done before. There is always something you can learn about this work. Blacksmithing, there's always something new to learn.

But unlike an apprentice, Philip knows that a new task can ultimately be completed.

I say, "Well I haven't done it before." They say, "Well could you do it now?" "Well," I say, "it's bending iron so I could do it." I go ahead and get my paper. Start drawing and there it is. It's new, but it's bending iron. It isn't wood.

Philip is open to the challenge of novel commissions because he is confident of his ability. That ability was formed by the traditional exchange of instruction from an aged master to a young apprentice and nurtured during the rigorous and disciplined performance of the craft according to those first precepts. The past, then, exerts a pleasant stabilizing influence on Philip, reminding him not just of his beginnings, but of his potential.

It would not be accurate to label Philip a romantic because of his affection for times past. He does not allow his thoughts to dally in yesterday's glories; rather, he uses the past as a resource. He sometimes visits the open-air Charleston market on Saturdays when a flea market is held. In the long, columned shed various hawkers offer old things, the kind of antique for which there is as yet no collector's frenzy. The helter-skelter assortment of objects reminds Philip of his years on Daniel Island and his first days in the blacksmith shop.

My mind went back to the old wagons and horse trace chains, tongues on wagons. It just take your mind back. You know what I be thinking? That someday I'll be putting some of those things together again. [You] look backward to see forward. To see forward that's the question.

If Philip ever retires he will not abandon his trade. He would like to remain active and indeed devote himself exclusively to making small tools and wrought-iron devices, what he sometimes calls "souvenirs" or "hobby work." In making these items he would further establish himself as the blacksmith he used to be.

Saturday mornings at the market remind Philip of his role in the community. The scattered implements from the past help him to recall his numerous accomplishments and moreover to recommit himself to his age-old profession. Philip is not a romantic; he is simply a person from the past. Immersed in old-fashioned ways from birth, he has rarely considered any other source of wisdom worth inquiry. Raised by his grandparents and trained by Peter Simmons, the values of the last century have exerted a strong influence on him. To this can be added the sense of antiquity which saturates the Charleston environment, where anything old is immediately regarded as intrinsically good. Thus, when teaching the traditional handwork techniques of blacksmithing, Philip observes, "It's a hard way, but a good way to know." Philip appreciates the past because it is useful to him. The tested precedents of collective communal experience, the traditions of his people and his craft, are the ultimate standard for important decisions. For Philip and other traditionally oriented people, it makes much more sense to trust what one already knows than to guess hopefully and blindly about a novel possibility.

Throughout his career Philip has continually brought the past into the present and pushed old-fashioned customs, sayings, ideas, and values into the future. Looking from the outside one might say that Simmons is preserving a tradition of skill and artistry against all odds, and that his effort is heroic. One could even conclude that a rough-hewn revolution providing an alternative to the mighty industrial combine is in progress. These conclusions would, however, be inaccurate. Philip acts as he does because that is the logical mandate of his training. He learned to be old-fashioned as a boy and had found that those old-fashioned ways of working have remained economically rewarding. But Philip favors the past not just because it is possible; he does so deliberately because he sees no better option. After all, "it's a good way to know."

Stepping out from his shop, Philip's eyes turn upward to the thick canopy of branches and leaves that shade the yard. He chants out loud, "Under the spreading chestnut tree," checks himself and starts again. "Under the spreading pecan tree, the village smithy stands." Having modified the line so that it more closely describes himself, blacksmith of Charleston, he then recites a few more stanzas of Longfellow's venerable poem. What other poem should a blacksmith know? He enjoys the images because they match his own viewpoint. At the end of the second verse, "For he owes not any man," Philip notes in an aside, "That's what I like, that's what would make me happy." He observes further on in the third stanza the truthful accuracy of the images of bellows and "heavy sledge."

Philip is the "village blacksmith." Indeed, he ministers to Charleston's needs as if it were a village. His business is all conducted with local folks whom he knows first hand. Clients learn of him by word of mouth. Philip does not advertise; he must be found in person if a job is to be discussed. He refuses to answer his telephone after eight o'clock in the morning, thus ensuring that prospective clients will more than likely walk up his driveway and around the side of the house to the shop. Business remains a personal, face-to-face affair. Agreements are sealed with a smile, a nod, and a handshake just like any village affair. The reputation of the craftsman is more binding than a signed document. Philip recognizes that he provides a significant service to the community. He embellishes the environment with delightful images, besides providing sturdy fences and gates to protect property. Customers depend on him and the ironwork he makes. Thus a bond is forged—the bond of responsibility for service rendered.

While the weight of this burden is lightened by local fame and prestige, Philip is quick to give credit to his customers. They are the ones, he says, who make it possible for him to work: "Without they asking me to do it, I can't do nothing." Considering further where the credit should best be placed, Philip contemplates his religious beliefs:

All work by man is the hand of God. Edison made the light, but everything he used made by God. Same for my work, I look at nature a lot. The greatest history is "In the beginning God created heaven and earth." Nothing before that; all comes from that, isn't it?

Philip Simmons is a humble man. He won't accept all the honor he is due. He wants mainly to serve others and in blacksmithing to "make a contribution to the city." He does not think that he has done enough; he never will think that.

Most think blacksmithing is a dying art, but it isn't. I feel confidence now and I know pretty good about iron, but I'm still learning, after

82 *Philip Simmons
    at his forge.*

fifty years. Fifty years ago and today—to me it's no different. I'm still heating and bending iron.

He is happy that he has apprentices to whom he can pass on his skills.

I take a lot of pride in my work. I do a good job. That's why I want my apprentices to learn well. I want them to do a good job.

Now that Philip is almost seventy years old, he is coming closer to taking the role Peter Simmons played.

Peter had brain power. Used that small hammer to point where to hit. Tink, tink, tink. Young man got muscles. I've got brain power. What I do now is thinking.

From the position of supervisor, Philip will direct the future of Charleston blacksmithing. His trainees will mature in the trade and will in turn draw more helpers. The craft will be passed from generation to generation. The current growth of local blacksmithing has both surprised and pleased Philip. Referring to Si and Ronnie he says, "These two boys, I didn't think they would still be at it now after they reached their thirty-second birthday. It's almost something that I didn't dream of." Commenting on the reasons for taking up this trade, Philip explains, "You got to love it to learn it." This is, in fact, what he has given his apprentices: a love of the craft. They too, are ready to challenge the future as blacksmiths.

In a chain running from a distant ancestor to Guy to Peter to Philip to his apprentices, the links are joined; the tradition of Afro-American ironwork is preserved. But beyond this personal passage of knowledge and technique, the role that blacks in Charleston have served since the first decades of the eighteenth century retains a communal significance for whites as well as blacks. The city of Charleston is often praised for its excellent ironwork, and consequently ornamental wrought iron has become a kind of civic emblem. We would do well to remember, however, that a portion of the fame, prestige, and esteem that is awarded to the city, in fact, belongs to the craftsmen. Their excellence has become the excellence of the community.

APPENDIX
The Blacksmith At Work

The Community and the Shop

Philip Simmons lives and works on the East Side of Charleston, one of several black neighborhoods in the city. It is bordered on its eastern edge by the harbor, and on the west by King Street and on the south by Calhoun Street. A random assortment of industrial and commercial enterprises crowds the northern end of the neighborhood. One of the most common buildings on the East Side is a two-story frame house (called a "single house" in Charleston) with its gable facing the street. These traditional structures are interspersed among contemporary housing projects. All the houses are packed tightly together and sit close to the sidewalk. Some buildings are in need of work, but many are well kept and have been recently painted. There is constant movement on the streets whether it is early in the morning or late at night—children at play, men and women coming from and going to work, old and young men jostling each other on the corner, folks running up to the corner store for a soda or over to the tavern for a beer.

One focal point of the neighborhood is C. A. Brown High School, a cluster of red brick buildings and asphalt playgrounds covering a full city block. Across the street from the school, a narrow driveway leads to a small gray building. This house at 30½ Blake Street is the home of Philip Simmons. The yard is filled with a brick driveway that winds around to the side of the house and back to the blacksmith shop. A few bushes make valiant attempts to grow in a flower bed without much success. Flat metal stock of various sizes is stored alongside the drive. The shop, roughly square in shape, is a metal-frame building covered with galvanized iron siding. It has a shed roof, sloping front to rear. To the right of the shop is a storage yard for scrap parts such as old gates, sections of fence, grates, and metal frames. Up against the wall of the shop an area is kept clear for assembling fence sections. On the left side of the shop factory-made bars in twenty-foot lengths are arranged in a covered rack.

The shop interior is kept neat and organized. The younger men who work with Philip sweep up, put the tools away, and generally keep the metal stock covered and in order. Tools are hung around the walls in their appropriate places. Chisels and punches go over the work table and vise. Tongs are hung above the forge and bending templates are kept along the back wall. Paint is kept on a shelf, while electric tools, drill bits, spare parts for oxygen-acetylene torches, and assorted small items are stored in a much-scarred wooden cabinet.

The items that take up floor space are also placed near the walls. The forge is in the rear right-hand corner with the anvil close to it. There are tables on both sides of the shop. The one near the forge is used for metalwork, usually to bend scrolls, while the other is used for storage. There is a vise mounted on a small platform and a bench grinder near the front right-hand corner. The left half of the shop is usually empty during the day but at night a low trailer, loaded with an electric welding machine and a cutting torch, is parked there.

Although the shop has two nine-foot-wide doors that can be swung open, only the left-hand door is used, and it is never closed. There are two windows, one on the east side and the other opposite it. The shop is fairly dark all the time, even on the brightest days, partly because of three shade trees which provide a kind of canopy over the building, and partly because there is no chimney to carry smoke from the forge out of the shop. Although a vent has been procured, Philip has not found time to install it. Consequently the ceiling and the walls are stained black by the soot. The darkness is intentional, since sunlight disguises the color of hot metal.

The two most important objects in the shop are the forge and the anvil. The forge is one that Philip made himself. It consists of a sheet-metal tub, twenty inches high and two feet in diameter, lined with bricks and fire clay so that a small opening remains in the center. An electrically driven fan forces air into the forge from the bottom. Philip has two anvils. One is used in the shop, while a smaller one (which once belonged to Peter Simmons) has been relegated to the scrap yard and is rarely used. The main anvil sits on top of a wooden box built in the shape of a truncated pyramid. The base is wide and the top conforms to the dimensions of the bottom of the anvil. The whole assemblage is not attached to the floor, so that the anvil can be shifted around to accommodate any particular size piece of iron.

Simmons has many tools (see list 1) but only a few are frequently used: a ball-peen hammer, a chisel (used to cut iron bars), the welder, the oxygen-acetylene torch, a forge rake (used to keep clinkers from clogging the forge), and a blacksmith stand (used to hold extra-long stock that projects over the edge of the forge). He has a set of tongs that were also used by Peter Simmons, but he prefers to work the iron in long sections. This way he can hold on to the piece while beating it with the hammer. Hand-holding the iron gives him a better "feel" of the blows so he can determine whether a piece needs to be reheated. His hands are heavily calloused and hence much less sensitive than the average person's to the burning heat. Sometimes he will gingerly reach into the forge and knock away chunks of molten slag with his bare hands, much to the fascination of onlooking visitors.

A Traditional Fire Tool

I once asked Philip if he could show me how he would have made a tool in the old-fashioned way. He answered me with a question: "Have you ever seen a log roller [a fire poker]?" When I replied that I hadn't, he drew pictures of three types. They were all very similar except for their different kinds of handles. One had a brass handle (figure 1A), while the other handles were fashioned from the end of the poker's shank. The simple ring was called a "poor man's handle" (figure 1B). Simmons said that his grandfather had owned one like that. The fancy spiral design was a "rich man's handle" (figure 1C). Simmons decided to make the poker with the rich man's handle because that would show more of the "old-time ways."

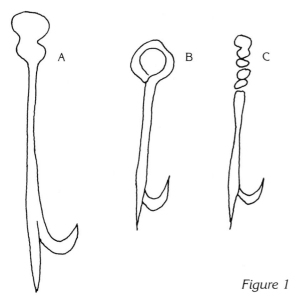

A
B
C

4. A welding flux made of borax and sand was sprinkled on the area.

5. When the area was white hot (melting temperature), the piece was removed from the forge and welded with hammer blows (figure 3).

Figure 3

6. The loop was cut at the end with hammer blows on a hardy (figure 4), a cutting tool which is inserted into a slot in the anvil (figure 5).

Figure 1

A good piece of wrought iron had to be selected first. Simmons wanted to use a piece of old Swedish iron because it could be worked more easily, it welded more efficiently, and it resisted corrosion better than American iron. In the yard he picked out an old gate which had a network of ⅜″ rods. He cut out a five-foot section and was ready to start:

1. A section of rod 4″ from one end was heated in the forge until red.

2. Four inches of the rod were bent into a loop with hammer blows (figure 2).

Figure 4

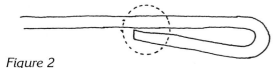

Figure 5

7. The two prongs were heated until red.

8. The poker was given its final shape. The welded prong was straightened and pointed with hammer blows. The continuous part of the rod was shaped into a hook by hammer blows over the horn of the anvil (figure 6).

Figure 2

3. The place where the end of the loop touched the rod (A in figure 2) was heated until red.

Figure 6

9. A 3″ portion of the rod 2′ from the handle end of the poker was heated until red.

10. The rod was placed in the vise and bent 90° to form an L such that 22″ was left to form the handle.

11. A 10″ section of the bent portion nearest the angle was heated until red.

12. The poker was placed in the vise next to a 1″ diameter rod.

13. The handle was wrapped around the 1″ rod three times with hammer blows (figure 7).

Figure 7

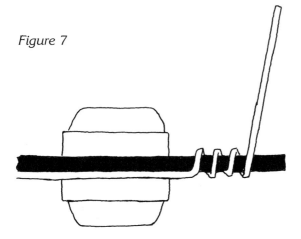

14. Ten inches of the unbent portion of the handle were heated until red.

15. The poker was placed in the vise next to a ¾″ diameter rod which was inserted through the previously formed coils.

16. The 10″ portion of rod was wrapped three times around the ¾″ rod with a small pair of tongs, yielding a handle represented in figure 8.

Figure 8

17. The whole handle was heated until red.

18. The coils were compressed with hammer blows and spaced over the hardy until the handle became, in Philip's terms, "egg-shaped."

19. The remaining bit of rod at the end was turned into a hook with hammer blows. The handle was then complete (figure 9).

Figure 9

The entire process took about twenty minutes. Most of the time was spent waiting for the iron to reach the appropriate temperature. The work itself was very short and quick. Philip told me that the poker should be exactly 36″ long when finished. He had done all the work—except for cutting out the original piece of iron—without the benefit of measure. However, when he checked the finished product it was exactly as he had predicted, 36″. His uncanny accuracy in the perception of lengths was further emphasized as he then sketched out in chalk on the floor a full-size fire tool of a different type. I placed the tool he had just made on top of the drawing and found it to be a perfect match in length and the placement and curvature of the hook.

Philip was happy with the finished tool. He said that he had not done a forge weld for a long time and was "rusty on his blacksmithing." But rusty or not his final product was an example of traditional craftsmanship at its best, combining function and beauty to make a tool well suited to its task. The hook, a gracefully curving prong, could easily get a secure bite on a burning log to pull it over efficiently. It was also flat enough so that if turned up it could be used to push logs back into the fire. The handle, though it might seem designed for elegance (as suggested by the name "rich man's handle"), was also made to be useful. Its elliptical shape fits the hand well. The hook in the end is a thoughtful convenience; it allows the tool to be easily hung away on a peg or nail. The pragmatic and aesthetic are closely bound in this object. However, Philip indicated that this piece was to be considered a tool and not an art work. He performed a brief pantomime to demonstrate the proper use of the poker. As he pulled and pushed at an imaginary fire he proved that this tool was indeed a log roller. It had been designed to do a specific job in a particular way.

Wrought-Iron Art

I was also present when Philip made a divider, a partition screen between a kitchen and a dining room. It was composed of four sections: a doorway panel, a pilaster post, a section that would rest on a counter top, and a panel surrounding a wall cabinet (figure 10). Philip asked the customer what kind of decoration she wanted, and he made a sketch on the spot. He later converted the sketch into a kind of plan. The area where the divider was to be installed had been thoroughly measured. These dimensions were carefully scaled out at ¾" to a foot, but the scroll and leaf work was simply drawn according to what looked right. The drawing process was tedious. Decorations were continuously drawn, erased, and redrawn. The paper was almost worn through when the plan was finished (figure 11).

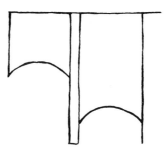

Figure 10

Figure 11

This drawing became a kind of blueprint which Philip constantly consulted as the work progressed. The crumpled and abused sketch was cast about the shop—hung on the wall, tossed on the floor—as the piece took shape. But it was never discarded; he kept it always close at hand. The drawing's importance was clearly demonstrated when it accidentally was left outside during a brief rain. Philip hastily retrieved it, but not before the sketch was thoroughly soaked. Although it looked like a lost cause, he carefully revived it. Delicately unfolding the paper, he laid it over some smoldering embers on the forge. When the drawing was dry and just about to burn, he snatched it up and went back to work.

The following steps were accomplished over a two-day period. There were constant interruptions and delays ranging from a tornado that knocked out the electrical power (the forge fan is electric) to running errands and answering the telephone. Some of the pieces were fabricated when I was not present, but I did see all of the parts assembled.

*   *   *

First a frame was made for the scrollwork in the doorway panel and pilaster post (figure 12).

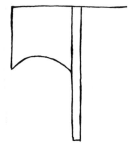

Figure 12

1. An 84¾″ section of a 20′ length of ½″ × ½″ bar was marked off for the top of the divider frame.
2. A wedge-shaped section was cut at that mark with the torch and the bar was bent 90°.
3. The angle was checked with a square and welded.
4. A 3′ length was measured off along the wall side of the doorway panel and cut with a torch.

5. The pilaster post was marked off on the top of the frame.
6. Two 6′ lengths of ½″ × ½″ bar were cut with the torch.
7. The two lengths were welded position 5″ apart.
8. A 43½″ length of ½″ × ½″ bar was cut to form the arch at the bottom of the door panel.
9. The arch was shaped on the anvil. It was curved to fit into a space 35″ wide.
10. The ends of the arch were cut with the torch so it would fit snugly against the vertical pieces (figure 13).

Figure 13

11. The arch was welded in place.
12. A 5″ piece of ³⁄₁₆″ × ½″ flat stock was cut for the bottom of the pilaster post with a chisel and welded in place.

*   *   *

The frame was laid on the floor of the shop and the pilaster post decoration was shaped (figure 14). This step involved forge work and bending iron on the work table.

13. An 8½′ length of ⅜″ diameter rod was cut for the pilaster post scroll.

14. The end was heated until red in the forge.

15. The end was turned back and beat into a 1″ diameter disk on the anvil.

16. Using notches on the side of the work table, the end was bent into a half-circle.

17. The pilaster design was chalked out on the floor inside the frame.

18. Working from the top down, each curve was bent and checked against the drawing on the floor.

19. The last half-circle at the bottom of the post was measured and the last 6″ of rod was cut off with a chisel.

Figure 14

20. The end was curled as in steps 15–17.

21. The whole piece was then welded to the frame. Simmons welded from the dining room side because the divider would be seen more from the kitchen.

*   *   *

The work then changed tempo as he began to make leaves. All the leaf decorations were made at one time. Two pieces of stock were worked simultaneously so that while one was being worked on the anvil the other could be heating on the forge.

22. The end of a ⅜″ diameter rod was heated until red and pointed with hammer blows (figure 15).

23. About 4″ of rod was heated until red and turned back with hammer blows (figure 16).

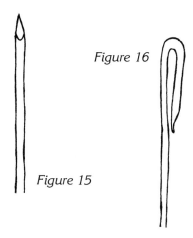

Figure 16

Figure 15

24. The loop was then closed by welding. (The arc welder was used here instead of forge welding because of problems with dirty coal. It wouldn't burn hot enough to get the iron up to welding heat.)

25. The whole loop was heated until red and then beaten on the anvil into leaf shape. The opening in the loop became a central spine on the leaf. The edge rather than the face of the hammer's striking surface was used

to raise a number of ridges resembling leaf veins (figure 17).

26. Steps 23–26 were repeated until twelve leaves were completed. (Eventually all but one were used on the divider.)

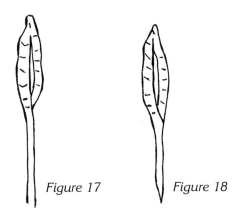

Figure 17   Figure 18

27. All the leaf stems were heated until red and pounded into a flat point to provide a good welding surface (figure 18).

28. A "sleeping leaf" was made following steps shown in figure 19.

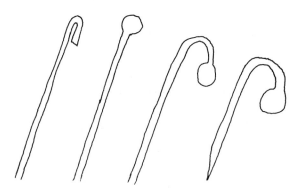

Figure 19

*   *   *

Next the scrolls for the doorway panel (figure 20) were fashioned with templates (see list 2). Again Simmons worked from the top down. All the scrolls were made from $^3/_{16}'' \times \frac{1}{2}''$ flat stock and worked cold.

Figure 20

29. Half of the scrolls were chalked out on the floor.

30. Sixty-three inches of stock were measured off for the first scroll. (Lengths are determined by bending the tape measure over the chalk lines.)

31. The metal was tapered on one end and given a tight curl with hammer blows. A special technique is used in tapering. The stock is first narrowed so that as it is drawn out the metal will spread back to its original width. If the metal were simply flattened, the scroll would have an uneven flare at the end (figure 21).

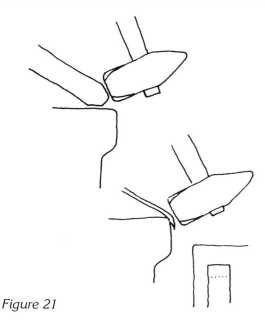

Figure 21

Figure 22

32. The piece was then curled around template number 2 and then matched to the drawing on the floor. When that scroll was correctly shaped it was used as a "sample" for all others of that diameter.

33. A companion was made for the first scroll following steps 30–32.

34. A second set of scrolls was made using 53" of stock.

35. A third set of scrolls was made employing template number 6 but with the same procedures and with pieces 24" long.

36. The last set was made from pieces 19" long using template number 8.

37. All the scrolls and three leaves were arranged in the doorway panel. Leaves were set in the pilaster post.

38. Leaf stems were bent on the anvil and the leaves were bent into what Philip calls a "wiggletail" (figure 22).

39. The shank on the center leaf of the doorway panel was extended. A length of ⅜" diameter rod was welded to the stem so the leaf could fill an empty space between the top set of scrolls.

40. All of the pieces were welded in place in the doorway panel and pilaster post (figure 23).

Figure 23

*   *   *

Work on the divider stopped here for the day. It was then ten o'clock. Simmons had by then put in a seventeen-hour day. I returned the next morning at nine to find that the rest of the frame had been completed.

41. A 6′ length of ½″ × ½″ bar was welded to the top of the frame to separate the counter and cabinet panels.

42. Forty-three and a half inches of ½″ × ½″ bar was shaped in an arch and welded to the bottom of the counter panel following steps 9–11.

*   *   *

The counter panel had ten scrolls. Four pairs were all the same (figure 24).

43. Intervals for the scrolls were marked off on the floor.

44. The first eight scrolls were shaped as in steps 32 and 33, using pieces 62″ long.

45. The last set of scrolls was shaped using 39″ lengths and template number 4.

46. Leaves were set out in a pattern similar to those in the doorway panel.

47. A long piece of ⅜″ diameter rod was attached to the stem of the center leaf so that it could touch the top of the frame.

*   *   *

Finally the cabinet panel was formed (figure 25). Since the scrolls in this section had to fit both against the wall and around a projecting cabinet, some tricky maneuvering was

Figure 24

Figure 25

required. The scrolls were set up with some "play" so they could be adjusted once the divider was in place.

48. The cabinet profile was chalked out on the floor.

49. Two scrolls were formed around template number 3 from pieces 50" long.

50. One scroll was made from a 19" piece with template number 3.

51. An **S** scroll was made using two templates. One end of the 68" piece of stock was curled on template number 1, and the other end was curled in the opposite direction on template number 5.

52. The leaves for the counter and cabinet panels were welded in place.

Figure 26

53. All of the parts for the two panels were welded in place. The divider was then essentially complete (figure 26). All that remained were a few decorative and finishing touches. All the welds were ground smooth and sealed with a glazing putty which "fulled the cracks." Bits of slag were filed and wire brushed away. Finally the whole piece was given two coats of antique gold paint at the request of the customer.

Conclusion

The main distinction between Philip's art and craft should be made on the basis of intention. In ornamental work he endures extensive anguish to "get it right." His repeated trials, tests, and debates are the expression of a struggle to mentally form a unique template for one particular art work. When he made the log roller, he worked quickly and assertively. There were no false starts, no dry runs. He simply selected a standard type and made the tool. The techniques of craft and art in ironwork are similar. Compare, for example, the first steps in making a leaf and a fire poker. It is therefore the marshalling of mental energy to create a beautiful object rather than a practical one that makes his room divider an art work.

List 1

Tool Inventory for Philip Simmons's Blacksmith Shop (1972)*

2 anvils

2 bench vises
5 C-clamps

3 ball-peen hammers
1 eight-pound sledge hammer
1 four-pound sledge hammer

8 Tongs: 2 flat tongs
        3 pick-up tongs
        1 hollow bit tongs
        1 pincer tongs
        1 oval iron tongs
        1 clip puller

1 box of anvil tools (swages,
  hardies, fullers, etc.)
15 bending templates (see Appendix B)

5 calipers
3 travelers
2 levels
2 squares
2 bevel squares
1 measuring tape

1 ladder

1 drawshave
1 wood saw
1 hack saw
1 hatchet
2 electric drills with wood bits

1 pipe cutter
2 chain cutters
3 bolt threaders
2 oxygen-acetylene torches
12 chisels
2 punches

1 forge rake
1 blacksmith stand

2 shovels
1 post-hole digger
2 pointed bars (5½')

1 bench grinder
1 portable grinder
1 file
1 rasp
2 wire brushes

1 arc welder

2 push brooms

*For analogs for these tools see H. R. Bradley Smith, *Blacksmiths' and Farriers' Tools at Shelburne Museum—A History of their Development from Forge to Factory*, Museum Pamphlet Series, no. 7 (Shelburne, Vt.: The Shelburne Museum, 1966).

List 2

Templates for Bending Scrolls

Type	Length*	Turns	Diameter
1. Half **C**	68″	2	15″
2. Half **C**	43″	1½	12″
3. Half **C**	38″	1	11″
4. Half **C**	40″	2	9″
5. Half **C**	34″	1⅓	7½″
6. Half **C**	12″	1½	6″
7. Half **C** (oblong)	32″	1	6″
8. Half **C**	17″	1	4½″
9. Half **C**	30″	1½	4″
10. Half **C**	27″	1¼	4″
11. Half **C**	13″	¾	3″
12. **C** (symmetrical)	34″	1	3¾″
13. **C** (asymmetrical)	1. 16″	1. 1	1. 3½″
	2. 18″	2. 1	2. 4″
14. **C** (asymmetrical)	1. 11″	1. 1	1. 1½″
	2. 12″	2. 1	2. 2½″
15. **S** 1. oblong	1. 18½″	1. 1	1. 5″
2. round	2. 4″	2. 1	2. 2¼″

*Lengths for templates with two sizes of scrolls are measured from midpoint of template to end of curl.

GLOSSARY

ANTHEMION FLOWER An ornamental design based on the honeysuckle flower, commonly used in Greek and Greek Revival architecture.

BROKEN SCROLL A pattern of decoration in which the flowing line of the curve is interrupted by a right angle.

DOG BARS A series of vertical bars closely spaced in the lower section of a gate to keep out stray animals. They are often tipped with barbed points.

DRIFT A tapered steel pin which is driven into a hole in a bar of iron with unmatched ends to align them properly.

FRETWORK A band of ornamentation formed from horizontal and vertical straight lines.

FULLER A blacksmith tool that fits into the anvil or is held on top of the heated iron. It is used to groove the metal or draw it out to thinner dimensions.

HARDY A chisel-shaped tool that fits into a slot in the anvil and is used to cut metal bars.

HEAT A term used to describe the period of time that an iron bar remains at the desired temperature; for example, "You should punch that hole in one heat."

HOLLOW TOOL A pair of tongs with jaws that form an open space when closed. It is most appropriate for holding large square section bars and pipes rather than flat bars.

LATCH BAR The section of a gate to which the locking mechanism is attached; usually a narrow horizontal band of decoration at hand level.

NEGATIVE SPACE The area immediately surrounding an object. The mass of the object is said to constitute positive space. A negative-space composition is brought into being and made interesting by the manipulation of material.

149

OVERTHROW	The top section of a gate, often arched or triangular in shape. It remains in a fixed position while the gate swings below it.
PALMETTE	A fan-shaped ornament composed of narrow divisions like a palm leaf.
PANEL	A section of the swinging portion of a gate.
PILASTER	A vertical rectangular element in wrought-iron design rigidly braced to support a gate or fence. Pilasters are usually placed on either side of the gate and at regular intervals along the length of a fence.
ROSETTE	A rose-shaped, flat, oval ornament formerly cast in iron, now in lead.
SCREEN	The section of a gate which is hinged and moveable.
SET	A type of hammer used to finish partly formed sections of the workpiece; so named because the tool *sets* into the final position when driven by a regular hammer.
STOCK	The basic iron and steel material as it is delivered from the foundry, usually in specific dimensions and lengths of twenty feet or more.
SWAGE	A type of tool used to change the shape of metal stock. The tool has two basic parts, top and bottom. The bottom swage fits into the anvil and the top swage is held over the iron. Swages can be used to draw a square bar down into a round rod.
UPSET	The process of making a bar shorter and thicker.
VOLUTE	The spiral end of a scroll.
WATER LEAF	A ribbonlike narrow leaf with crinkled edges. It is usually aligned with the lower edge of a decorative panel, but turned up at the end.

BIBLIOGRAPHICAL NOTES

While this book is about the life of a single blacksmith and celebrates a single individual's achievements, it may also give rise to questions about his environment, his community, and his profession. The reader may want to have more background on all of these subjects. This information may be provided in books and articles which are cited in this essay. The sources mentioned do not exhaust the available bibliography, but are mainly those works that I found helpful in telling Philip Simmons's story. Each section below corresponds with a chapter and includes the works used to write that chapter. Although some sources were employed as references throughout, those are only cited where they were first used.

The Sea Island Experience

No study of the black experience in South Carolina can fail to mention the outstanding work of Peter H. Wood, *Black Majority: Negroes in Colonial South Carolina from 1670 through the Stono Rebellion* (New York: W. W. Norton, 1974). This book makes the case for a viable black culture in the Carolina low country. Wood not only shows connections to Africa, but also investigates the intimate ties of the Sea Islands with the Caribbean, a theme pursued by Richard S. Dunn in an article "The English Sugar Islands and the Founding of South Carolina," *The South Carolina Historical Magazine* 72 (1977): 81–93.

The distinctive African qualities of Sea Island life

were, of course, highlighted earlier by Melville J. Herskovits in *The Myth of the Negro Past* (Boston: Beacon, 1958), in which he declared the doctrine of African cultural survivals in the Americas. Herskovits drew upon a series of sociological writings in making his manifesto. The coastal islands near Beaufort were the focus of a number of studies, including Guy Johnson, *Folk Culture on St. Helena Island, South Carolina* (Chapel Hill: University of North Carolina Press, 1930); Guion Griffis Johnson, *A Social History of the Sea Islands* (Chapel Hill: University of North Carolina Press, 1930); T. J. Woofter, *Black Yeomanry: Life on St. Helena Island* (New York: Henry Holt, 1930); and Mason Crum, *Gullah: Negro Life in the Carolina Sea Islands* (Durham: Duke University Press, 1940).

Given this specific focus of scholarly attention, the islanders remained the subject of academic inquiry even when they left their homes. Clyde Vernon Kiser traced their movements in *Sea Island to City: A Study of St. Helena Islanders in Harlem and Other Urban Centers* (New York: Columbia University Press, 1932). The Penn Center on St. Helena Island, a vocational training school and social service center for blacks, provided a base of operation for visiting scholars. One of these visitors was photographer Leigh Richmond Miner. His sensitive pictorial record of the activities of the center and daily life in surrounding area circa 1900 is preserved in Edith Dabb's *Face of an Island* (Columbia, S. C.: R. L. Bryan, 1970).

Other island communities are encountered in a set of popular rather than scholarly works: Charlotte Kaminski Prevost and Effie Leland Wilder, *Pawley's Island: A Living Legend* (Columbia: State Print Co., 1972); Nell S. Graydon, *Tales of Edisto* (Columbia, S. C.: R. L. Bryan, 1955); James Henry Rice, Jr., *Glories of the Carolina Coast* (Columbia, S. C.: R. L. Bryan, 1925); and Guy Carawan and Candie Carawan, *Ain't You Got a Right to the Tree of Life?* (New York: Simon and Schuster, 1966). Works which give some information on agricultural practices in the general vicinity of

Daniel Island, where Philip Simmons grew up, include John B. Irving, *A Day on the Cooper River* (Columbia, S. C.: R. L. Bryan, 1932); Arney R. Childs, ed., *Rice Planter and Sportsman: The Recollections of Jacob Motte Alston 1821–1909* (Columbia: University of South Carolina Press, 1955); Archibald Rutledge, *Home by the River* (Indianapolis: Bobbs-Merrill, 1941); and Patience Pennington, *A Woman Rice Planter* (Cambridge, Mass.: Harvard University Press, 1961).

Recent interest among historical archaeologists in the textures of black life has led to excavations of Sea Island plantation sites. These efforts in turn have led to a reevaluation of the foodways of slaves; see John Solomon Otto, "A New Look at Slave Life," *Natural History* 88 (1977): 8–32. A relatively recent dissertation by Mary Arnold Twining, "An Examination of African Retentions in the Folk Culture of the South Carolina and Georgia Sea Islands," (Indiana University, 1977), contains data collected from Johns Island. While reviewing the whole issue of African cultural transfers in the local crafts, she includes important descriptions of net knitting.

Black Craftsmen and Blacksmithing

The history of black craftsmanship is inextricably tied to the rise of the free black class. A fine general work on this dimension of American social history is Ira Berlin, *Slaves without Masters: The Free Negro in the Ante-Bellum South* (New York: Vintage Books, 1974). Sources that deal more specifically with South Carolina are Marina Wikramanayake, *A World in Shadow: The Free Black in Ante-Bellum South Carolina* (Columbia: University of South Carolina Press, 1973); Ulrich B. Phillip, "The Slave Labor Problem in the Charleston District," *Political Science Quarterly* 22 (1907): 416–39, and E. Horace Fitchett's two articles, "The Traditions of the Free Negro in Charleston, South Carolina," *Journal of Negro*

History 25 (1940): 139–52 and "Origin and Growth of the Free Negro Population of Charleston, South Carolina," *Journal of Negro History* 26 (1941): 421–37.

The role of black workmen in American history is reviewed by Charles H. Wesley in *Negro Labor in the United States 1850–1925* (New York: Vanguard Press, 1927); black working conditions in South Carolina are presented in George Brown Tindall's *South Carolina Negroes 1877–1900* (Columbia: University of South Carolina Press, 1952); and the early phases of the history of Charleston's black craftsmen are touched upon by Richard Walsh in *Charleston's Sons of Liberty: A Study of the Artisans* (Columbia: University of South Carolina Press, 1959). Two works that assess the extent of black skills in the twentieth century are W. E. B. Dubois, *The Negro Artisan* (Atlanta: Atlanta University Press, 1902) and W. E. B. Dubois and Augustus Granville Dill, *The Negro American Artisan* (Atlanta: Atlanta University Press, 1912). In all of the works cited above, blacksmiths are specifically mentioned as noteworthy craftsmen. For a broader survey of black artisans consult Robert Farris Thompson, "African Influence on the Art of the United States," in *Black Studies and the University,* edited by Armstead L. Robinson et al. (New Haven: Yale University Press, 1969), pp. 128–77; and the first sections of James Porter, *Modern Negro Art* (New York: Dryden, 1943) and David C. Driskell, *Two Centuries of Black American Art* (New York: Alfred A. Knopf, 1976).

For extensive discussions of the specific aspects of blacksmithing, such as the names and use of tools, processes of metalworking, properties of different metals, and techniques of forging, see M. T. Richardson, *Practical Blacksmithing* (New York: Weathervane Books, 1958); Alexander G. Weygers, *The Modern Blacksmith* (New York: Van Nostrand Reinhold, 1974); Aldren A. Watson, *The Village Blacksmith* (New York: Thomas Crowell, 1968); Alex W. Bealer, *The Art of Blacksmithing* (New York: Funk and Wagnalls, 1969); and Jack Andrews, *Edge of the Anvil: A Resource Book for the Blacksmith* (Emmaus, Pa.: Rodale Press, 1977). In chapter 2 I reconstruct some of the tasks Philip Simmons learned to perform. The information contained in these works describes the widest range of smithery, much of which Philip has no experience in, even though he probably could do most of the jobs listed.

Charleston Ironwork and Philip Simmons

The examples of Philip's work presented in this volume obviously exist amidst a set of older ironworks; his pieces derive content from them and at the same time contribute to the local tradition. The best historical study of Charleston decorative wrought iron continues to be Alston Deas, *The Early Iron Work of Charleston* (Columbia, S. C.: Bostick and Thornley, 1941). Since Deas's book does not consider works done after the first decades of the nineteenth century, the reader must turn to studies of local architecture to gain a more complete survey of Charleston's decorative iron. The following works give descriptions of gates and balconies as well as fancy homes and public buildings: Alice R. Huger-Smith and D. E. Huger-Smith, *The Dwelling Houses of Charleston* (New York: J. B. Lippincott, 1917); Elizabeth Gibbon Curtis, *Gateways and Doorways of Charleston, South Carolina in the Eighteenth and Nineteenth Centuries* (New York: Architectural Book Publishing Co., 1926); Albert Simons and Samuel Lapham, eds., *Charleston, South Carolina* (New York: Press of the American Institute of Architects, 1927); and Samuel Gaillard Stoney, *This Is Charleston* (Charleston: The Carolina Art Association, 1944).

Susan and Michael Southwick's recent *Ornamental Ironwork* (Boston: Godine, 1978) contains a brief chapter on Charleston, but more important, provides comparative examples of ironwork from other American cities. Consulting Ifor Edwards, *Davis Brothers, Gatesmiths* (Cardiff: Welsh Arts Council, 1977) or John Starkie Gardner,

English Ironwork of the Seventeenth and Eighteenth Centuries (London: B. T. Batsford, 1911) will give a good sense of the British and Welsh antecedents of Charleston's ornamental wrought iron.

A few articles in newspapers and magazines provide slight glimpses of Philip Simmons's work: Charlotte L. McCrady, "Modern Blacksmith Uses Ancient Anvil," *Charleston News and Courier*, 5 September 1955; Otis Perkins, "Blacksmith Loves His Ancient Art," *Charleston News and Courier*, 19 November 1957; Ann W. Dibble, "Local Ironsmith Still Plies Trade," *Charleston News and Courier*, 27 January 1974; Jack Leland, "Flying Sparks Called Simmons to Anvil," *Charleston Evening Post*, 2 July 1976; and Carol Speight, "The Creative Hand," *South Carolina Wildlife*, July–August 1977, pp. 24–41, esp. pp. 28–29.

Aesthetic Considerations

The material that furnishes the basis for chapter 4 is taken primarily from transcripts of interviews with Philip Simmons. Readers may, however, be interested in other analyses of the personal element in a traditional expression. Michael Owen Jones in *The Handmade Object and Its Maker* (Berkeley: University of California Press, 1975) reviews at length the creativity of a Kentucky chairmaker. An earlier study by Ruth Bunzel, *The Pueblo Potter* (New York: Columbia University Press, 1929) has one chapter entitled "The Personal Element in Design."

My attempt to put Philip Simmons's standards in his own terms reflects current trends in social science toward regarding native terminology as preferable. For another study with similar motivation see D. M. Warren and J. Kweku Andrews, *An Ethnoscientific Approach to Akan Arts and Aesthetics*, Institute for the Study of Human Issues, Working Papers in the Traditional Arts, no. 3 (Philadelphia, 1977). Another elicitation of native standards is found in Robert Farris Thompson's "Yoruba Aesthetic Criticism," in *The Traditional*

Artist in African Societies, edited by Warren d'Azevedo (Bloomington: Indiana University Press, 1974). It is interesting to observe that Philip Simmons's relative categories for density and curvature are matched by relative Yoruba categories for straightness, delicacy, and spacing. Another important study of African creativity rich in specific details about craftsmen's inspirations is Warren d'Azevedo's article "Sources of Gola Artistry" in the book cited just above.

A few Afro-American folk artists have been studied, and their opinions are recorded in the following books and articles: William R. Ferris, Jr., "Vision in Afro-American Folk Art: The Sculpture of James Thomas," *Journal of American Folklore* 88 (1975): 115–31; Elinor Lander Horwitz, *Contemporary American Folk Artists* (New York: J. B. Lippincott, 1975); Worth Long and Roland Freeman, "Leon Rucker: Woodcarver," in *Black People and Their Culture: Selected Writings from the African Diaspora*, edited by Linn Shapiro (Washington: Smithsonian Institution, 1976), pp. 33–34; and Lynda Roscoe, "James Hampton's Throne," in *Naïves and Visionaries,* edited by Martin Friedman (Minneapolis: Walker Art Center, 1974), pp. 13–20. The theme of improvisation that accompanies the spirit of prophetic vision sensed in all the works described in these articles is encountered in other traditional black art forms, namely rhetoric and music. For a discussion of the aesthetics of black innovation see Daniel J. Crowley, *I Could Talk Old-Story Good: Creativity in Bahamian Folklore* (Berkeley: University of California Press, 1966); Roger D. Abrahams, *Deep Down in the Jungle: Negro Narrative Folklore from the Streets of Philadelphia*, 1st rev. ed. (Chicago: Aldine, 1970); Bruce A. Rosenberg, *The Art of the American Folk Preacher* (New York: Oxford University Press, 1970); and Charles Kiel, *Urban Blues* (Chicago: University of Chicago Press, 1969).

Different Roles

The brief account of Philip Simmons's family life which I have given reveals a pattern of extensive care and concern. The network of inclusive relationships and responsibilities is presented in greater detail by Carol B. Stack, "The Kindred of Viola Jackson: Residence and Family Organization of an Urban Black Family," in *Afro-American Anthropology: Contemporary Perspectives*, edited by Norman Whitten and John Szwed (New York: Free Press, 1970), pp. 303–12, and in her book *All Our Kin: Strategies for Survival in a Black Community* (New York: Harper and Row, 1975).

Philip's account of his baseball-playing days may raise questions about the experiences of semiprofessional black ball players in the years of segregation. For a similar account from West Virginia see Reginald Miller's "Conversations with the 'Ole Man': The Life and Times of a Black Appalachian Coal Miner," *Goldenseal* 5 (1979): 58–64. The best personal material on black baseball is to be found in Theodore Rosengarten's "Reading the Hops: Recollections of Lorenzo Piper Davis and the Negro Baseball League," *Southern Exposure* 5, nos. 2–3 (1977): 62–79.

Miscellaneous

My research on Philip Simmons has resulted in several publications of related interest. These are "Philip Simmons: Afro-American Blacksmith" in *Black People and Their Culture*, edited by Linn Shapiro (Washington: Smithsonian Institution, 1976), 35–57; *The Afro-American Tradition in Decorative Arts* (Cleveland: Cleveland Museum of Art Press, 1978), esp. pp. 108–21; and "The Craftsman and the Communal Image: Philip Simmons, Charleston Blacksmith," *Family Heritage* 2 (1979): 14–19.